ADVANCED DREDGING TECHNIQUES

PART 2

The Right Approach

By

Dave McCracken

For more information about the author and his ongoing programs, either visit www.goldgold.com or write to him at :

Dave McCracken
27 Davis Road/P.O. Box 47
Happy Camp, CA 96039
www.goldgold.com

Layout & Design by Maria McCracken

Published by
Keene Industries
20201 Bahama Street, Chatsworth, CA 91311
(818) 993-0411
www.keeneeng.com

Table of Contents

Acknowledgement

The production of this book was a team effort. I would like to acknowledge the tireless contributions from several of my very close friends:

Copy editing was meticulously performed by Scott Langston. English composition was not one of my stronger subjects in school! As Scott could only work with the material I provided him, he certainly invested a lot of personal effort here. The book is a lot better because of that.

Doyen Salsig enthusiastically did the final proofing and editing.

My wife Maria, as always, was there from beginning to the very end. She scanned and gathered up all of the raw material from my earlier writings on the subjects covered here and made them available for me to work with. After Scott and I had finished our part, Maria typeset the final material into the finished product that is in your hands.

The creative material comes from me. But, it involved a lot of hard work from Scott, Doyen and Maria to put it all into final form. And, for that, I am truly grateful.

Dedication

This book is dedicated to George Massey.

Since gold was discovered in California during 1848, no single person has ever even come close to doing more to promote and institutionalize small-scale gold mining in America than the "Buzzard." He is truly the father of recreational gold mining. We all benefit from his hard work.

Introduction

During the past several years, I have had the opportunity to train hundreds of people in basic gold-mining techniques, as well as several dozen men and women in commercial, underwater-mining programs. I have also had several occasions to observe many dozens of folks conduct their own mining operations in Africa, South & Central America, Alaska, Canada, Indonesia, Cambodia, Madagascar, and on our own rivers of Northern California.

Running a successful mining operation is one thing. Training someone else to run their own successful mining operation is different altogether. Because, in addition to working with mining theories, principles, and procedures, we are also now working with a student's ability, or lack of ability, to apply these values to actual situations in the field.

I know a fair number of successful gold miners. Some of them I trained; some learned on their own. Many are successful on a recreational scale. A few of them mine gold to support themselves and their families.

I also know a number of rather unsuccessful miners, both commercial and recreational. We trained some of those, too. Some of the others would not have accepted help to save their lives.

Unquestionably, there is a distinct difference between successful miners, partially successful miners, and those who are completely unsuccessful. The difference is seen in the basic approach they use. A simple way to understand the differences is to use the analogy of radio wavelengths.

Picture each person as an electronic, radio-frequency tuner, living in a universe that has an infinite number of possible radio frequencies that one could tune in to. I have noticed that successful gold miners are more often tuned to a particular "frequency" than those who are not so successful.

The unsuccessful ones were, more often than not, unreceptive students. How do you account for someone who seeks out training, but won't listen to the instruction? How do you explain a person who obviously needs help, but when it is given, they: want to argue, refuse to understand, become suspicious, and reject the training offered? Helpful information is being broadcast, but the "student" is not tuned to the "frequency" so the signal can be received.

One of the common characteristics I often see in successful commercial and recreational miners is a compelling drive to get into, and to stay on, a pay streak during their mining activities. It's like a gnawing hunger. You can actually see the intensity of their desire in everything they do. This is the same kind of drive that you will find in an entrepreneur searching for a good investment, a business person scrapping to lock up a profitable transaction, a used-car dealer closing on the sale of a car, a musician creating a magical melody, or the competitive desire of an athlete to win.

All gold miners would like to be successful and find a lot of gold. The difference is that successful miners actually **create** their own success. They not only learn how to do it, but they hustle to find the best opportunities for applying their mining knowledge in the field.

Learning the lessons in this book will only enhance the larger chemistry that is required. The successful ones will **not allow** themselves to fail. Their intense desire to succeed provides them with the hunger, the focus to learn. And it carries them beyond, driving them to seek out the right opportunities to apply their energies and knowledge. In a sense, successful gold prospectors have contributed to their own prosperity by tuning in, and staying tuned, to a **learning** wavelength.

I look back on my early involvement with gymnastics in high school. I was moderately successful--enough so to become co-captain of our team during my senior year. But, there were others I competed against who were far better than I was. I look at the gymnasts of today, the ones who seem to be near perfection, and I realize that I was **never** in their league. Why? At that time, I felt that those who were better than I just had more inherent gymnastics ability than I did. But, the truth is: they were more focused and determined to achieve gymnastics perfection than I was. By their commitment and hard work, they **made** themselves into better gymnasts.

There is no right or wrong about this. You simply have a better chance of becoming exactly what you focus your creative energies upon.

A more "sensitive" observer may say that I lacked the proper coaching. But that's just an excuse. It does nothing to help a person "dial in the required frequency."

My wife's son, Derek Parra, wanted to become the world's fastest speed roller skater. So he finished high school, with honors, a half year early so he could get started sooner. Then, with no money or financial support, he moved to Florida where he could be near a world-class coach. He took a part-time job at McDonald's for the first year. He followed his coach to Maryland and then to Delware where he worked part-time at the roller rink. His duties ranged from managing, being a DJ, to doing maintenance work--whatever it took to prove his commitment. As a result, he made the United States Roller Skating Team during his first year, took a Gold Medal at the World Championships, and placed 1st in the U.S. National Competition shortly thereafter. That success did not come without cost. He worked **really hard!**

Over the next few years, Derek became one of the world's best-known in-line skaters, and he acquired several sponsors that were paying him pretty well. But, when he realized in-line roller skating was not going to be an Olympic sport during his time, Derek made the difficult decision to give it all up, move to Wisconsin, and take a job near an ice-skating rink. There, he started all over again. In a close competition, he earned his way onto the U.S. Speed Ice Skating Team. At first, the competition was overwhelming, and he finished his races far behind the winners. But he never gave up. Slowly, but steadily, he overcame every obstacle that kept him from winning races. He was always tuned to the wavelength of success. And, he poured all of his creative energy and commitment into becoming the fastest ice skater he could be. During the 2002 Winter Olympics in Salt Lake City, as a complete dark horse, he took a Gold Medal in the 1500m race and a Silver Medal in the 5000m for America, breaking two World Records along the way! That had to be the most exciting thing I will see in this lifetime!

Becoming really good at something requires focus and dedication. It means tuning into the right frequency that will lead you to everything you need to know and do to achieve success. A person has to **want** to be on that frequency in order to learn how to do something well.

If you would like to be successful at gold mining, dial **yourself** into that frequency. It can be found inside of you. The key is true desire and personal commitment!

Author's Note: One thing I have witnessed over the past 20 years, is that a lot of women have taken a keen interest in gold prospecting and have become quite good at it. I have had the pleasure, and honor, to work along side a number of female gold dredgers who really knew what they were doing with a suction nozzle! It is heartening to see this evolution in the gold-mining community, as it broadens into a fun pursuit that the whole family can enjoy. Especially because of this reality, this book is intended for the use and benefit of the female members of your family, just as much as the males. With this understanding, I want to make it clear that any reference to a male pronoun in this book, implies and includes the female as well. While both are specifically mentioned together at times, a single pronoun reference is often made to smooth the flow of the text. This is not intended as a slight in any way. We love our women folk! And, we are grateful for their increasing interest in sharing a passion and a pastime that has so traditionally been male-dominated. May God bless them all!

Happy hunting,

Dave McCracken

1

PROSPECTING FOR GOLD IN HARD-PACKED STREAMBEDS

Successful placer mining is generally accomplished in two steps: (1) prospecting and (2) production. This is true on any scale of operation. First, you need to find a gold deposit--usually a pay streak. Once the deposit is located, then you can concentrate on a production operation to recover the gold from the deposit.

Prospecting generally consists of digging or dredging sample pits/holes in different locations, then comparing the results of the different tests. In a waterway, the first effort should be to locate the common path along which most of the gold typically travels. Sampling is then performed along that path in order to find the pay streaks. This system is thoroughly outlined in *Gold Mining In The 21st Century*, and Part 1 of *Advanced Dredging Techniques.*

When you are sampling, you should always be

looking for hard-packed streambed material. "Hard-pack" is created at the bottom of waterways during major flood storms.

Generally, a winter storm, even a **large** winter storm, will not create enough turbulence and force in a river, creek, or stream to redeposit the streambeds that are already in place along the bottom.

Flood storms of the magnitude to redeposit streambeds do not occur often.

For example, on the Klamath River in northern California, where I conduct a lot of my own dredging operations, we believe that the last time a new, hard-packed streambed was formed was during the 1964 flood. In most of the channel along the Klamath, the 1964 flood layer was laid down on top of a much older, harder-packed, virgin streambed--one formed perhaps thousands of years ago.

So, it takes a **major** flood storm to move and lay down a hard-packed streambed. And, it takes a super **major** flood storm to create enough force and turbulence in a river to break up ancient streambeds and redeposit them as new, hard-packed streambeds along the course of the waterway. This happens only very rarely.

The reason that hard-pack is important to a prospector is because gold nearly always concentrates at the bottom of hard-packed, flood layers. Therefore, it is important for a prospector to target his or her sampling efforts to reach the bottom of hard-packed streambeds.

Gold is about six times heavier, by volume, than

the average weight of the sand, silt, and rocks that are associated with gold in a streambed. Because of this disparity in weight, when streambed material is being washed down river in a major flood storm, most of the gold will quickly work its way down to the bottom of the streambed material that is being carried along by the raging flood waters.

Because the gold is so much heavier, it will work its way down the river channel more slowly than the other streambed materials. During major flood storms, most of the gold moving in a waterway will be washed down along the bedrock or across the surface of hard-packed streambed that is not being moved by the storm. At some point during the storm, gold becomes trapped out of the turbulent flow by dropping into irregularities, cracks and holes that are present along the surface over which it is traveling.

It is reasonable to assume, because of its weight, that gold generally deposits much earlier in the course of the storm than the streambed material that eventually lays in on top of a pay streak. As the storm begins to taper off, and the water forces start to slow down, particles of gold will begin to drop out of the flow. Along the path where most of the gold is traveling in the waterway, traces will be deposited, with more substantial pay streaks being dropped in the larger, low-velocity areas. These gold deposits are being laid down even as the (much-lighter) rocks and gravel are still being washed down river. Streambeds form later in the storm, when the water turbulence tapers off enough to allow the rocks, gravel, sand, and silt to drop out of the flow and seat themselves along the bottom.

The streambed material that lays in on top of a pay streak will nearly always be hard-packed. The reason for this is that if there is enough force and turbulence to move substantial amounts of gold in the waterway, then there is also enough force to create a naturally-formed streambed on top of the gold as the storm tapers off.

How do natural streambeds form? First, the storm needs to be powerful enough to rip up the streambed material, put it into suspension, and wash it down the waterway. On a much larger scale, this process is similar to gravel washing through a sluice box. During the later course of the storm, when the water force and turbulence starts slowing down, natural obstructions or traps along the river bottom will allow certain key rocks to become lodged or seated. An example might be a flat rock that drops into a bedrock indentation, with the forward edge of the rock pointing slightly downward into the flow. The water current then holds the rock to the bottom. With this rock in place, new locations are formed for other rocks to become lodged. Smaller rocks, sand, gravel, and silt will fill every gap and crevice in a manner similar to the mortar used to cement layers of brick. More rocks then fall into place. More filler then packs the spaces created between the rocks, and the bed forms.

Hard-packed streambeds form mostly with the flat rocks lying horizontally and slightly tipped downward in the direction of the current.

The way in which streambeds form during major flood storms leaves the bed material seated and compacted together, much like a mechanical structure. In dredging, we call this structure "hard-pack." Most of the gold in present waterways is covered over by hard-packed streambed.

For substantial amounts of gold to move in a waterway, it requires a major flood storm with enough power to **blast** up the existing streambeds. It is the abrasive (more like a huge saw) action of the streambed material, as it is being washed down along the bedrock during major flood storms, that causes bedrock channels to be cut deeper into the earth over geologic time. When such a storm tapers off, new hard-packed streambeds will form over top of the gold deposits.

"Loose-packed" material consists of sand, silt, rocks, and gravel which contains little or no natural structural cohesiveness. When you dredge a hole down through loose streambed material, it tends to keep sliding in on you. Digging through loose-pack can be like trying to dig down through a gravel or sand pile. The material keeps slipping into the hole. This makes for a much more difficult sample hole if the material runs deep.

Sometimes, loose material is resting on top of a hard-packed streambed, and must be moved out of the way first, before the hard-pack can be properly sampled.

When you dig or dredge a test hole through hard-pack, the streambed will generally hold up the wall surrounding the hole. In fact, many of the old-time operations would tunnel underneath hard-pack. This was/is called drift mining.

It is also important to be able to identify tailings. Tailings are easy to spot as long as you understand how streambeds are mined. Tailings do not look either like hard-pack or loose streambed material.

When hard-packed streambed is being mined, the cobbles and boulders (i.e., rocks that are too large to pass through the recovery system) are tossed back onto a pile behind the production area. As the production area moves forward, piles of boulders and cobbles are left behind, in place of the original hard-packed streambed. Sometimes, the sand, silt, and gravel that is processed through the recovery system is dumped on top of the cobbles. Later, winter storms also wash sand, silt, and gravel across the top of the cobbles. The sand, silt, and light gravel then filters down and fills in most of the space between the cobbles. Therefore, tailings usually end up as loose stacks of cobbles with sand, silt, or light gravel filling the spaces.

A large flood storm may wash the cobbles away at a later time, and redeposit them into a newly-formed hard-packed streambed. Or, as in the case of the 1964 flood in northern California, rather than wash away all of the cobbles, many of the piles were leveled off and a new layer of hard-packed streambed was deposited on top of the tailings. When we dredge sample holes in those locations, we usually find a hard-packed streambed on top of the tailings. If gold traveled in that part of the waterway, we find it concentrated at the bottom of the hard-pack, sitting on top of the tailings. Underneath, we find loose cobbles with sand and silt between them. These usually go all the way to bedrock. We find very little gold on bedrock because it has already been mined.

Some waterways will have several different natural streambed layers, each with its own concentration of gold resting on top of the layer or bedrock below. Different streambed layers usually exhibit a different color and compactness. The contact zone between the layers is generally pretty easy to spot. On the contact zones between layers, and on top of bedrock, is where you will find most of the gold concentrations. **Those areas, then, should be the focus of your sample holes**.

Seldom will you find rich, pay-streak gold deposits associated with loose streambed material. However, it does occasionally happen. These occurrences are almost always the result of winter storms, and the related run-off, eroding away a hard-packed streambed along the bank--which washes the gold down into the waterway to rest with the loose material.

Effective sampling is the key to a successful mining operation. And, when you are sampling, you should be looking for hard-packed streambed layers. Watch for the gold concentrations along the bottom of these layers, because that's where you will usually find them.

2

PRODUCTION GOLD DREDGING

Dredging technique in gold-bearing streambeds is centered mainly around removing the oversized rocks (i.e., any rock which is too large to be sucked up the suction nozzle) from the work area of the dredge hole. In most hard-packed streambeds, this removal of the oversized material is the bulk of the dredger's work. From a production standpoint, a large portion of this work has to do with freeing and removing the oversized rocks from the streambed in an orderly fashion.

Hard-packed streambeds are laid in horizontally, in layers, during major flood storms. Generally, the streambed is put together like a puzzle, with different rocks locking other rocks into place. Usually, because of gravity, rocks higher up in the streambed (laid down later) are locking lower rocks (laid down earlier) into their position in the streambed.

Because of this, we have found the best means of

production is to dredge the hole down a layer at a time. If you take down a broad horizontal area of the streambed together, you uncover a whole strata of rocks which are interconnected like a puzzle. Then, you can see which rocks must be removed first in order to free the others more easily. This really is the key to production dredging!

Generally, there are two things that drastically slow down the production of less-experienced gold dredgers: (1) plug-ups and (2) "nitpicking."

The time and energy spent freeing plug-ups from the hose and power-jet cuts directly into how much progress will be made in the dredge hole.

Plug-ups are caused by rocks jamming in the dredge's suction hose or power jet. Everyone gets a few of these. Inexperienced dredgers get many! This comes from not understanding which types of rocks, or combination of rocks, to avoid sucking up the nozzle. Basically, this knowledge comes from hard-won experience in knocking out hundreds of plug-ups to the point where you are more careful about what goes into the nozzle. I have covered this area quite thoroughly in my book, *"Basics of Successful Gold Dredging."*

During the past several years, the various dredge manufacturers have started putting out dredges using oversized power jets. This means that the inside diameter of the jet tube is no longer reduced in size from the inside diameter of the suction hose, helping to greatly reduce the number of plug-ups a dredger will get while in production. So, the combination of some practice in learning which types of rocks to avoid sucking up the nozzle, and the oversized power jet on modern dredges, will eliminate most of the plug-ups that would otherwise hinder production.

NITPICKING

The other main loss of production, which we are all guilty of to some extent, comes from a poor production technique which we call "nitpicking." Nitpicking is when you are trying to free rocks out of the streambed which are not ready to come out.

Nitpicking is dredging around and around rocks which are locked in by other rocks that need to be freed up first.

When you find yourself making little progress, the key is almost always to widen your dredge hole.

Production dredging means moving gravel through the nozzle at optimum speed. This is accomplished by making your hole wide enough to allow the oversized rocks to be easily removed. As soon as you find the rocks are too tight to come apart easily, it is usually time to widen the hole again.

DREDGING HELPERS

An underwater dredging helper can be a **big** help in removing oversized material from the dredge hole.

Basically, there are two different types of jobs in an underwater mining operation: (1) nozzle operator (N/O) and (2) rock person (R/P). The nozzle operator is responsible for getting as much material up the nozzle as possible. Therefore, it is his or her responsibility to be in charge of how the dredge hole is taken apart. The rock person has the responsibility of helping the nozzle operator by removing the rocks that are in the way of production.

Since there are plenty of oversized rocks that need to be removed from a production dredging operation's work area, it is important that the R/P have some judgment as to which rocks ought to be

The rock person should be moving the very next rock that is in the nozzle's way.

moved first. Ideally, he will pay attention to what the N/O is doing, and focus his efforts on those rocks which are immediately in the nozzle's way.

Silt is usually released when a rock is moved out of the streambed. So a R/P must be careful to not cloud out the hole (loss of visibility). This can be avoided by concentrating on (a) the rocks that are near to where the N/O is working, or (b) the rocks that have already been freed from the streambed, but not yet removed from the hole.

On occasion, a R/P unwittingly takes over the production operation by randomly moving whatever rocks he or she happens to see. This, then, causes the N/O to have to follow the R/P around, sucking up the silt which would otherwise cloud out the hole. This generally results in slowing down production. It is better if the R/P accepts the position of being the N/O's assistant, which allows the latter to direct the progress of the dredge hole.

The nozzle-operator should communicate with his helpers where he wants to take a cut off the front of the hole.

Since production is controlled by how efficiently rocks can be removed from the dredge hole, it is important to understand that they must also be strategically discarded in a manner which will not require them to be moved a second or third time. Generally, this means getting each rock well out of the hole as far back as necessary. Good judgment is necessary on this point. You only have so much time and energy available. How far that oversized rock will need to be removed from the hole is somewhat dependent on how deep the hole is going to go. Early on, we often must guess about this. Sometimes, the streambed turns out to be deeper than we thought it would be. Then, we find ourselves turning around and throwing all the rocks further back away from the hole. This is normal.

We also must use judgment to not waste our limited time and energy resources. We don't really want to move all those tons and tons of rocks any further than necessary to allow a safe, orderly progression in the dredge hole.

As we move our hole forward, and as we dredge layers off the front of the hole, we try to leave a taper to prevent rocks from rolling in on us. This is an important safety factor! Also, since the N/O's attention is generally focused on looking for gold, the R/P should be extra vigilant in watching out for safety concerns. Any rocks or boulders that potentially could roll in and injure a teammate should be removed before they have a chance to do so.

Some rocks will be too large and heavy to throw out of the hole. Therefore it is good technique to leave a tapered path in the back of your hole for use in rolling boulders up and out. If you can't remove boulders from your hole, the hole may become bound with oversized material. This can create a nitpicking situation.

If you see boulders being uncovered which are too large to roll out of the hole, you can make room for them at the back of your hole, on the bedrock, by throwing other rocks further behind. This takes a little advanced planning.

The rock-person's attention should always be on safety. If there is danger, he or she should point it out to the nozzle-operator.

Sometimes winching is necessary to remove large boulders, or an abundance of large rocks, from a dredge hole. However, because winching takes time, the pay streak has to be richer to justify the time and effort that would be required.

The amount of streambed material that you are able to process through a gold dredge will determine the volume of gold that you will recover. Actually, this volume-to-recovery ratio is true of any type of mining operation, whether it is a large-scale lode mine, a recreational miner using a gold pan, or anything in between.

The smaller the amount of material that a dredging program has the capacity to process, the richer the pay dirt must be to recover the same amount of gold. Consequently, a smaller-volume operation needs to invest more time and energy in sampling, to find the scarcer, higher-grade pay streaks. For this reason, smaller-volume operations generally spend more time sampling, less time in production, and, therefore, usually recover less gold.

However, successful gold recovery depends on more than the size of the dredge. **A lot** has to do with the skill of the operator(s).

In summary, then, it can be said that, as a general rule, gold mining, on any scale, is volume sensitive. If you can dredge twice the volume of streambed material, not only can you recover twice the amount of gold, but you may find (many) more lower-grade gold deposits which can help you meet your minimum paying standards.

This is why I always advise beginning gold dredgers to go out and find an easy location, where they can practice their basic gold dredge production techniques to improve their speed, **before** they start a serious sampling program.

A beginner will sometimes be so slow in volume production that he or she may miss valuable pay streaks simply for lack of being able to process enough gravel, or making his sample holes large or deep enough during sampling.

When you dredge a sample hole, you must evaluate how much gold you recover against the amount of time and work it took to complete the test hole. If you are only moving at 20% of your potential production speed, you can make the mistake of walking away from excellent pay streaks just because you will believe they are not paying well enough.

It is a good idea to practice your dredging skills so that you can operate the nozzle efficiently while moving your own oversize material out of the way.

When we run larger-sized (8" or larger) gold dredges, we almost always have at least two divers working together underwater. The reason for this is because running an eight-inch dredge in six feet or more of streambed material requires that a large number of oversized rocks must be moved out of the dredge hole by hand. This varies from one location to the next. But, generally, in hard-packed, natural streambeds, somewhere between 60 and 85 percent of the material is too large to process through an eight-inch dredge. This is where the second person becomes a **big** help. A sole operator in this type of material, when it is deeper than five or six feet, is going to spend a great deal of time throwing rocks out of the hole, rather than operating the suction nozzle.

The smaller the dredge, the shallower the material that a single operator will be able to manage efficiently. Also, some hard-packed streambeds require that most of the oversized rocks be broken free with the use of a pry bar. This further decreases the amount of nozzle time on a single-person dredging operation.

In the final analysis, it is the volume of material that is sucked up the nozzle (in any given location) that determines gold production. However, the efficiency with which the oversized material is moved out of the way has a direct impact on how much gravel and gold will be moved through the nozzle. So, gold production ultimately is controlled by how well the oversized material (rocks too large to pass into the suction nozzle) is managed. The N/O's focus, therefore, should be on directing the nozzle to suck up the gravel that will make it easier to free more oversized rocks, rather than indiscriminately sucking up any gravel in the hole.

If a R/P is added to the operation, he or she must increase the program's efficiency by at least as much as the percentage of gold which they are going to receive. This is not difficult to accomplish if conditions are right and the dredgers are organized.

On the other hand, if you are operating a dredge in two or three feet of hard-packed streambed, adding a second person may not increase your production speed enough to make it worthwhile to pay them for their time. This is because the material is shallow, and you may not have to toss the oversized rocks very far behind in the dredge hole.

When I am operating a production dredge in five or six feet (or more) of streambed material, I can literally bury a rock person with oversized rocks, and make my helper work like an animal all day long. I also have to work like an animal to accomplish this. The result is a good-paying job for my helper and a substantial increase in my own gold recovery.

When I find myself working in eight, ten, or more, feet of material, I **must** have at least one or two rock person(s) to help me. Otherwise, I will be completely buried with cobbles and oversized rocks all day long and will get in very little nozzle time.

All of this also applies to smaller-dredge operations. Your ultimate success will be directly proportional to how much material you can get up the nozzle of your dredge in the right locations. The more efficiently you can make that happen, the more quickly you will get your nozzle into pay dirt.

Large-rock management is important! Since **most** of the material in a natural, hard-packed streambed will be too large to go through the suction nozzle, the progress and speed of your operation will directly depend on how quickly and efficiently the oversize (cobbles and boulders) materials are moved out of the N/O's way. Dredging is **not** just a matter of sucking up some gravel, at least not in the places where most pay streaks are found. In the pay streaks, the only gravel that gets sucked up is that which is found between the oversized rocks.

So there is a lot of good to be said about organizing a system to get the rocks moved out of your way as quickly and efficiently as possible. This is true in dredging on any scale and regardless of whether you are working alone or with a team.

The reason that larger dredges can get more accomplished in gold-bearing streambeds is usually not because they will move more gravel. It is because a larger hose and nozzle will suck up a bigger rock. Every rock that can go up the nozzle is a rock that does not need to be packed out of the dredge hole by hand. This is the primary factor that speeds things up with a larger dredge.

Another major advantage of a two-person team is the substantial amount of emotional support which a second person can add to the operation-especially when you are dredging in deep material, or when you are sampling around for deposits and haven't found any in a while.

On the other hand, the **wrong** person can inhibit the operation, either physically or emotionally. So you must be especially careful to select teammates who have work, moral, and emotional drives that are similar to your own.

In my own operations, we have found that the key to good teamwork is in establishing standard operating procedures for almost everything we do. We also work out standard underwater signals. This takes quite a bit of planning and communication in advance, and it is an ever-continuing process. We have standard procedures for removing plug-ups from the hose and jet. We have standard procedures for moving the dredges forward and backward during operation. And we have standard procedures for every facet of the underwater work-- from moving the raw material that is in front of us, to the tailings and cobbles behind our dredge hole.

As noted above, **volume** is the key to success-or, at least, to the **degree** of success. We take this aspect of the work quite seriously in my own operations--to the point where every single second and every single physical effort is regarded as important to the operational moment while we are dredging. You will never find the members of my team socializing or goofing off during the underwater production hours. We goof off a lot during the rest of the time. But during production time, we are entirely focused on the needs of the operation. We treat the dredging portion of the operation like a competitive team athletic sport. We don't compete against each other. We compete, in a team effort, against the barriers that "Mother Nature" has constructed for us to overcome in order to recover volume amounts of gold. To us, our teamwork is kind of like running a relay race. It is fun this way, when we really synchronize our efforts.

We try to spend a minimum of six hours doing production dredging or sampling each day. In our operation, this is normally done in two three-hour dives. Other commercial operators prefer three or four shorter dives. I know one commercial dredger in New Zealand who prefers a straight six, seven or eight- hour dive. What an animal!

I like lunch. But I do agree with the concept of long dives. The reason being, it takes a little while to get a good momentum going underwater. Every time you take a break, you need to get that momentum going all over again. What do I mean by "momentum"? Momentum in dredging is very similar to the beat of the drum in driving music. It is the continuous flow of gravel up the nozzle, with the oversized rocks being moved out of the way in their proper order, at just the right time, so that the flow of material into the nozzle is never slowed down. It is everyone doing whatever is necessary to keep that flow happening. It's about finding the right rhythm and keeping the beat alive!

In fact, team dredging can be kind of like an art form, similar to playing music. Only instead of notes being played on several instruments to form a harmonious melody, the team produces a stream of coordinated effort, using their bodies to move the suction nozzle and the oversized rocks, in concert, so that all of the effort works together to achieve optimum momentum and harmony.

In their initial enthusiasm, an inexperienced R/P

Four-man team on author's commercial operation. Notice how everyone is involved up where the action is happening at the suction nozzle.

may, rather than move the next rock in the path of the N/O, instead move the wrong rock and cloud out the dredge hole with silt. This causes the N/O to be slowed down (1) because of the decreased visibility, and (2) because he or she must now take the time to move the proper rock out of the way. This is the equivalent of "playing off key," or playing the wrong musical selection. Everybody else is playing one song, while the new guy is doing something entirely different. The bottom line for your dredging operation: less volume through the suction nozzle.

On the other hand, there is enormous personal and team satisfaction to operating within a well-structured, team-dredging system. Such a system functions best when the N/O is the conductor, and the R/P plays a supporting role, anticipating in advance and taking every possible action to contribute to the N/O's momentum.

This is not just about the next rock which is in the way of the nozzle, but also about moving the dredge forward a bit when necessary, to give the N/O a little more suction hose when it is needed. It is also about the dozen or so other things that are necessary to keep the flow going.

Every time the N/O has to put the nozzle down to take care of something, volume momentum is lost. That will impact directly upon the success of the operation. Every effort should be made to keep this "down time" to a minimum.

Since volume is **really** the key, on my operations, we try to treat dredging pretty much like a team sport - competitive and physical! When I give my R/P the plug-up signal, he races to the surface to quickly clear the obstruction. He doesn't just **mosey** on up there. He **moves** up there, like running for a touchdown or rounding the bases for a home run. This is also the sense of urgency with which he returns to the hole, once the plug-up is free. When he sees that rocks are stacking up in the hole, he doubles his pace to catch up. When caught up, he looks around to see where other cobbles might be moved out of the way, without clouding the hole. Or he might grab the pry bar and start breaking rocks free for me. At the same time, I am doing my own job-getting as much material through the nozzle as humanly possible while freeing new rocks to be moved by the R/P, with the minimum number of plug-ups. And I don't stop for anything if I can help it. If something else needs to be done, I delegate it to my R/P, or other helper(s), so that I can keep pumping gravel up the nozzle. Everyone's gold income depends directly upon how much material is sucked up.

While this statement is also true when you work alone, **every** effort counts for something in production team dredging. Therefore, everyone needs to pay attention to what is going on in the dredge hole. R/P's particularly have to be able to switch gears quickly. At one point, there may be a pile of rocks which needs to be thrown out of the way. The next moment, even before the R/P has moved many of those rocks, he or she may notice something else that is directly impeding the N/O's progress-like a boulder that needs to be rolled out of the way, or a particularly difficult cobble which needs to be broken free with the pry bar. The main objective in everyone's mind should be to **support the N/O's progress**. Whatever impedes his progress should get **immediate** attention.

The only time I intentionally slow things down is when I am uncovering the gold. I have to keep an eye on that to follow the pay streak. If the streak is good, I also point out the gold to my helpers as I uncover it. There is emotional gain from this. Everyone deserves the boost. The gold eventually gets spent. The memories last forever.

The key to a productive underwater support person is having him or her work to help keeping gravel flowing into the suction nozzle.

When things get too confused inside the dredge hole, sometimes the N/O must put down the nozzle and help organize in the hole. But everyone should be keenly aware of the fact that **actual production** momentum has **stopped** and that it needs to get underway again as soon as possible.

As noted earlier, we take cuts off the front of the dredge hole in production dredging, and take the material down to the bedrock in layers. We do this because it is the fastest, safest, and most organized method of production dredging.

Sometimes, when conditions are right for it, a R/P may be working right next to the nozzle, breaking the next rock free and quickly throwing it behind in the hole. However, on every cut, there comes a time when the N/O decides to drop back and begin a new cut to take off the next layer. The R/P has to pay close attention to this and follow the N/O's lead. Otherwise, he'll finish breaking free a rock up in the front of the hole when there is no nozzle there to suck up the silt.

In other words, the R/P has to keep one eye on what the N/O is doing all the time. Because if the N/O is a dynamic and energetic sort, he or she certainly will not be following the R/P around the dredge hole. The N/O has to orchestrate the effort. So, all of the support players should be paying close attention and assisting in the N/O's next move. A good nozzle operator takes apart the hole with an organized method that is easy to follow by the others.

The teamwork even extends up to the dredge tender on the surface, if you have one. The dredge tender should always monitor the water volume flowing through the sluice box. If it visibly slows down, he should suspect a plug-up and take steps to locate and clear it. Sometimes, the water volume has been reduced simply because the N/O has temporarily set the nozzle down over a large rock in the hole. But on the occasions when there **is** a plug-up, it is a mark of great teamwork to have a tender handling the problem, immediately, without having to be told. Volume through the sluice box should also be continually on the tender's mind. When gravel stops flowing, he should also be thinking that something is wrong.

That same level of anticipation and teamwork should also apply in other areas, as well. When the tender sees that the dredgers are moving forward in the hole, he should ensure that the dredge is being moved forward accordingly, so that the N/O will always have a comfortable amount of suction hose to work with. Good teamwork involves a lot of close observation and timely anticipation in order to minimize the number of actual orders that need to be given and the consequent amount of down time. Most of the activity is handled by standard operating procedures which require a bit of planning and communication in advance.

As you would imagine, there are many different opinions about all of this. Some people are simply not running any races. This is fine, but they must understand that they do not have the same gold recovery potential as others who are working at a faster pace--or with a more organized system.

Generally, you won't hear anyone on my team complaining--especially during split-up time.

There isn't anything difficult to understand about successful gold dredging techniques. The process is quite simple. Serious dredging on any scale is a lot of hard work. Volume is the game. The faster, deeper and more efficiently you can dredge the sample holes, the faster you will find the pay streaks--and, the better you will make them pay. Even with smaller dredges.

At the times when you are not finding commercial amounts of gold, there is at least the satisfaction of knowing that you are accomplishing optimum performance. And, when you do locate the deposits, the sky is the limit!

3

FAST WATER DREDGING

Author waves from where he is dredging in fast water.

Something we have known for quite some time is that pay streaks, often very rich pay streaks, exist in the fast water. At first, this may seem contradictory to our general understanding that gold pay streaks form in areas of the waterway where the water slows down.

However, we must keep in mind that pay streaks are created during major floods. During a major flood, a sudden drop in the bedrock can cause a very good gold trap, like the riffles in a sluice box, but on a very large scale.

If you turn on a garden hose at slow speed, the fast-water area is found directly where the water flows out of the hose. But, when you turn the water pressure up, the water momentum forces the water farther out. This condition also occurs in a river during a major flood. Areas where the water runs fast during low-water periods are more likely to be drop zones for gold during high water. The heavy momentum/velocity area will be forced farther downstream, leaving a drop zone for gold just below the bedrock drop. This explains why you can often find pay streaks under rapids when the river is flowing at summer levels. It also explains why you seldom find pay streaks within the first slow-water area below a set of rapids when the river is running at low levels.

Another reason why you may find gold in fast water is because dredging fast water is more difficult. Therefore, others are less likely to have mined there before you - including the old timers. For this reason, fast-water areas can often be virgin territory!

What exactly is "fast water"? This depends on each individual. It's primarily a matter of the diver's comfort level. To some people, if the water is moving at all, it's too fast to dredge. Other dredgers are able to dredge in water moving so fast, that the air bubbles created by the turbulence eliminate all visibility. After diving in really turbulent water, a person's equilibrium can become so disoriented that he/she can hardly stand up without weaving, as if intoxicated.

Several years ago, a friend and I were operating a five-inch dredge in some very fast, shallow water. Because of the extreme turbulence, one of us would work the nozzle, while the other one would hold onto the dredge to keep it from flipping over. The water was so swift, my friend was swept out of the dredge hole time after time. Once, he was carried away so fast, he didn't have time to untangle himself from his air line before he reached the end of it. The air line was tangled around his neck! There he was, flopping around in the current, like a flag snapping in a stiff breeze, tethered by the air line around his neck and struggling, unsuccessfully, to regain his footing in three feet of water. After he got safely to the bank, we were both laughing so hard that tears were streaming down our faces. I admit we were a little bit crazy. Fifteen minutes later, I was the one bouncing in the current behind the dredge, facing backwards at the end of an air line caught between my legs. Needless to say, my friend thought this was a riot! Dredging in fast water can be fun and exciting (not to mention the gold you can find). But, you must be aware of, and prepared for, the dangers involved. If you get into a situation that is beyond your ability to handle, there is very little margin for error.

SAFETY

Notwithstanding all the excitement and gold, safety should **always** be the most important personal consideration. **You** are the one out there in the field with the responsibility for using good judgment about what you can safely do, without cutting your margin for error too close. The river does **not** have any sympathy for people who "get in over their heads." I've known several dredgers who lost their lives by over-stepping their personal safety boundaries. It only takes a single mistake. The rest can happen very quickly. All the gold in the world is not worth dying for.

For the sake of safety, it makes good sense for you to not dredge in water that is faster than you are comfortable with. You will have to decide what that is. It is best to practice first in slower water, to gain experience and confidence.

One important thing you should remember about working under water: everything may be calm and under control right now, but five seconds later, you can find yourself in the most life threatening emergency you have ever experienced! This is even true in slow water. But, fast water gives you much less margin for safety if you make an error. You should not dredge in fast water if you are unable to control the various problem-situations that could develop.

You need to anticipate each problem that could possibly arise and work out your response, in advance.

Contrary to what many people believe, being swept down river by the current is **not** the major concern. This is a normal happening in fast water dredging. As long as you have your mask clear and your regulator in your mouth, being swept down river by the current is generally no big deal. That is, of course, unless you are dredging directly above a set of falls or extremely fast water.

In most cases, the "fast water" you are in is not a steady flow of current. It is usually turbulent, varying in direction and intensity. A swirl can hit you from the side and knock you off balance. Or sometimes it can even hit you from underneath and lift you out of the hole and into the faster flow. If you get swept down river in fast water, you usually just need to grab hold of the river bottom and work your way over to the slower water, nearer the bank. This movement is best done by continuing to face upstream, into the current, while you point your head and upper body towards the river bottom. That posture will drive you to the bottom where you can get a handhold on rocks or cobbles to anchor yourself down. Then you can work your way upstream, through the more slack current near the bank, and back out to your worksite again. This is all pretty routine in fast-water dredging.

Getting a hole **started** is one of the most difficult aspects of fast-water dredging. Once the suction nozzle gets a hole started (the suction nozzle in the hole can serve as an anchor to help hold you there against the current), and there are several cobbles behind you to use as footholds, it gets a little easier to hold a position there. After the hole has been expanded to the point where you can get your body inside it, you will find significant relief from the effects of the current's flow. But, it can sometimes be a real challenge until you do get to that point! At times, you may find it necessary to start your hole in slower water, then gradually work your way out into the faster current.

One of the main concerns when dredging in fast water is having your mask and/or your regulator swept or knocked off your face. This situation is one that can cause a person to panic, especially when both mask (vision) and regulator (air) are lost at the same time.

PANIC

There isn't a person among us who won't panic, given the right situation. People who say they will never panic under **any** circumstances are just not facing reality and, obviously, have never come close to drowning. I believe it is better to understand and acknowledge your limitations **before** you get into trouble. The closer you cut it on safety issues, the more aware of your limitations you should be. And, the more important it is to plan in advance how you will react to certain anticipated emergencies. It's too late to make such plans **after** something bad happens!

For me, it takes a lot of personal discipline to stay under control when an unexpected rush of turbulent water jerks my mask off and drags me, blindly and chaotically, down river. This has happened to me on several occasions. I know that under those circumstances, it would not take much more confusion (e.g., air line getting snagged, my body being banged against something, losing my balance, getting a breath full of water from my regulator, etc...) for me to totally lose control and freak out (panic).

I've worked with some who have a higher tolerance for panic in the water than I do. And, I know others who feel panicky as soon as they put their heads under water, even under perfectly-controlled conditions. We are all different, and we each have our own particular point at which we will panic in different circumstances. Everyone has a limit. It is better that we **not** delude ourselves about this. If you allow yourself to get overly "macho," and continually put yourself into situations that can take you beyond your limit, you will most certainly find yourself tested in a life or death situation sooner or later.

Panic is a survival mechanism that takes over when your mind is convinced that your life is in grave danger. At this point, your animal instincts take charge and deprive your intellect of the ability to reason things out. Panic tells you that there is no time left, that you are literally fighting for life just before unconsciousness. The situation **demands** that you spend your last/maximum physical effort to remove yourself from the danger that may be about to mortally injure you or cause you to lose your life. Panic is a horrible, terrifying, and, sometimes, embarrassing experience that happens when your normal, rational self totally loses control, and the animal part of you takes over.

There are milder versions of panic. Someone might "panic" and do something silly or foolish in a business or a personal setting. That's not the type of panic I'm talking about here. I'm talking about the **raw** panic that grips you at the moment you realize you may be at the point of losing your life.

There's always a chance of getting into serious trouble any time you are working under the water. Trouble under water is serious because humans cannot breathe water. There is no margin. You are either breathing air or you are not. It is an **immediate** emergency when there is no air. And, such emergencies can happen in a split second, **any time** you are under the water.

TAKING EXTRA PRECAUTIONS

Underwater vulnerabilities are especially present during dredging activity. Some of this vulnerability is due to the fact that it is necessary to weigh yourself down with lead weights in order to stay on the river bottom. Extra weight is needed to give you the necessary stability and leverage to control the suction hose and nozzle and to move rocks and obstacles out of your way. The demands of dredging activity require divers to be so heavily weighted down, you couldn't possibly swim to the surface without first discarding the weights that hold you to the bottom.

One of the most serious dangers to a dredger is the possibility of being pinned to the bottom by a heavy rock or boulder. All of the oversized rocks that can't be sucked through the dredge nozzle must be moved out of the hole by hand or with the use of winching equipment. When undercutting the streambed, or taking apart the dredge hole, there is the possibility of larger rocks rolling in on top of you. This possibility increases when you are working in turbulent, fast water. The erratic changes in the pressure that the water exerts on the exposed streambed material, inside and around the dredge hole, can cause boulders to loosen up and roll into the hole. These same boulders, if located in a streambed where the water is running more slowly, might not loosen up the same way, if at all. For this reason, a fast water dredger must take **extra** precautions to remove **all** larger sized rocks when they are exposed. One of our mottos is: *"You have to get the boulders before they have a chance to get you!"*

When working in fast water, all of your normal safety precautions, preventative maintenance measures, and common sense instincts must be scrupulously observed. Fast water may be thought of as a liquid flow of energy that is constantly challenging you and your equipment. Murphy's Law *("anything that can go wrong, will go wrong")* is always at work in fast water. It is hard enough to deal with the things you cannot anticipate will happen. You will have enough of these as it is. But, if you neglect to take action with respect to those things that you **can** reasonably expect to go wrong, you will almost certainly fail in your efforts to dredge in fast water. If it is wrong, *fix it now*, before it gets worse!

OPERATIONAL CONSIDERATIONS

In my experience, I have found that it is physically possible for me to dredge in water that is too fast for the safety of my dredge--even the kind of dredge that is best suited for fast water. Therefore, the need to operate in an environment that is safe for your dredge is one of the major limiting factors in fast-water dredging.

Most fast water dredgers add more flotation to their dredge platforms to give more stability. This can be done in many different ways, including additional pontoons, inflated tire inner tubes, PVC pipe material, Styrofoam, etc.

One of the main considerations when adding more flotation to a dredge is to avoid increasing the drag against the current. Additional drag causes problems in two ways: (1) the fast water current puts more strain on your dredge, frame, and tie off lines; and (2) more importantly, the surface tension caused by all that additional water dragging around the dredge makes it difficult to work near the dredge when you're in the water (which can be a particular problem when you're trying to knock out plug ups from the suction hose near the dredge).

Another goal when adding flotation is to keep the floats as narrow as possible. A wide set of floats is more likely to be tossed or dragged around by the turbulent flow of fast water.

Generally, when working in fast water, I try to find a location for the dredge where the water is a bit slower, just next to the fast water where I plan to work. This way, I can enter the river in slower water and work my way out underneath the faster water.

Otherwise, if I position the dredge directly in the fast water, it will become necessary for the divers to contend with fast water when entering the water from the dredge. This can be done, but it just makes the operation more difficult--especially, when the dredgers need to get back onto the dredge. Also, the buildup of cobbles and tailings near the dredge can add to the surface tension and create an even faster current flow under and around the dredge.

In the type of setup where the dredge is positioned off to the side in slower water, your suction hose will be running perpendicular, at least to some degree, to the flow of the fast water. That much hose exposed broadside to the current creates enormous drag, which can cause the suction hose to kink at the point where it attaches to your suction jet. Hose kinks will cause continuous plug-up problems. Therefore, you may find it necessary to disconnect the suction hose and cut off the section that has been kinked. However, you cannot shorten your suction hose very much before you lose the amount of operational flexibility you need for freedom of movement while dredging.

Suction-hose kinks can usually be avoided by setting up a special harness to support the hose in fast water. This is often done by rigging one or two extra ropes down from your main tie-off line. The ropes are fastened to the suction hose at points which will allow the hose to be flexed back by the current, but not to the critical kinking point. You **must** allow the hose to flex back. It is the bend in the suction hose which allows you the movement to expand the size of your dredge hole.

It is best, when rigging a fast water harness, to rig it in conjunction with your main dredge tie off line. This way, the entire dredge and suction-hose harness will move together, as a unit, when you take the equipment forward as your dredge hole progresses.

Larger, commercial dredges may be equipped with booms, which can be extended out in front and used to secure a suction-hose safety harness. In this manner, when the dredge moves forward, the suction-hose safety harness moves with it, as in the situation above. These booms are being used quite successfully for this purpose.

Another concern in fast-water dredging is to keep your suction nozzle and hose from being swept out of your dredge hole. Sometimes, the current will put so much drag on the suction hose that it takes all of your strength and energy to get any nozzle work done at all! In such a case, you can relieve the main strain of the drag by tying a section of the suction hose to a large rock at the rear of the dredge hole. When doing this, always leave enough slack in the hose to allow you to move the suction nozzle forward as your dredge hole progresses. Also, be sure to remember to untie the suction hose from the river bottom before you move the dredge. Otherwise, you can damage the hose by causing kinks in the middle!

When you take a lunch break or knock off for the day, you can anchor your hose and nozzle by either piling rocks on the suction nozzle or tying the nozzle to a large rock in the bottom of the dredge hole. It's no fun to start a production dive by having to work against the current to get your suction hose back into your dredge hole, because the fast water blew it out after your last dive. But, of course, all fast water dredgers get many chances to experience this; it's a normal part of the routine!

One safety point: when using ropes under water, don't use any more than is absolutely necessary. Always cut off any excess rope or pile rocks on top of it to hold it down. If there is a length of loose rope flopping around in the current, something (like your air line) always seems to get tangled in it. Loose rope under water is dangerous!

Your air line can be another source of problems when dredging in fast water. Always be sure to get all the loops out of your air line before starting your dive. Otherwise, the current can pull these loops into kinks, which can immediately cut off your air supply.

When you turn around in your dredge hole, to roll boulders, toss cobbles, or do any of the many other things associated with production dredging, get into the habit of exactly reversing your turn when you face forward again (turning back counterclockwise is "cancelled out" by turning forward clockwise). This practice will help prevent you from putting lots of loops in your air line during the course of the dive. Each loop is a potential kink that can cut off your air supply in fast water. Each loop also increases the amount of drag being brought to bear on your air line in fast water.

If you should get a kink in your air line that cuts off your air supply, you can usually get some immediate relief by pulling your air line in toward your body and letting it go. When you let it go, the pressure is temporarily removed from the kink, and you can usually get a breath of air. I always try this once, quickly, as soon as my air cuts off. If that doesn't give me immediate relief, I head straight for the surface so I can properly correct the problem.

If you are experiencing **any** difficulty with a kinking air line, your best course of action is to immediately remove every single loop in the line. Getting rid of the loops will require you to rotate yourself in circles, going in the appropriate direction, until the air line is straight again. Several years ago, I was dredging in fast water with a guy who had to repeatedly dive out of our dredge hole because of a kinking air line. After about the fifth time, I suggested that he take the time to straighten out his air line to fix this problem (I was beginning to worry that he would never think of doing this himself!). This remedy only worked for a short time, because he had developed the bad habit of turning around and around in the dredge hole as he was moving rocks, which just created more and more loops in his line. Fifteen minutes later, he was diving right back out of the dredge hole again.

Today, you can buy a heavier type of "safety" airline that will prevent kinking in all but the fastest of fast water. I recommend this heavier air line to anyone who plans to dredge in swift current.

By the way, your air line is also your direct connection to the dredge and to safety. When you connect your air line to the dredge, even in slow water, it should be wrapped around the dredge frame several times before being attached to the air fitting on the dredge. Most air fittings are made of brass. If you should need to use your air line to pull yourself to the dredge in an emergency, it is better that you not have to depend on the strength of a brass fitting!

Nearly all experienced dredgers are aware of the fact that their air lines are an extension of themselves while under water. Especially in fast water, it is **very** important that you not allow your air line to tangle around parts of the dredge, underwater obstacles, and/or the air lines of other divers in the dredge hole. If you cross **over** the top of another diver's air line, keep that in mind, so you will be sure to cross back **over** it again when you return. Each time you go to the surface, to remove a plug up or for whatever reason, take a moment to untangle your line from anything it may have wrapped around. As a practice, all dredgers should always untangle their air lines each time they return to the surface after a dive.

One of the persistent problems of dredging in fast water is the heavy drag on your air line. This can normally be solved by pulling some slack line into the dredge hole and anchoring it with a single cobble placed on top. This will allow some slack air line between you and the cobble. You want to be sure that your cobble anchor is not so large that you can't quickly free your air line in an emergency. Also, when you leave the dredge hole, don't forget to disconnect your air line from your anchor first.

Full face masks are generally not well suited for diving in swift water. Since they are larger, with substantially more surface area, they are more likely to get accidentally dislodged from your face. This can happen when the mask is bumped on another diver, or an obstacle, or when turbulent water catches it, especially from the side. To further complicate matters, when a full face mask fills with water, the regulator usually does as well. Having to clear the water out of your mask and regulator at the same time can be more difficult and contribute to a panic situation. I personally find that I am more prone to feeling panicky when something goes wrong inside of a full face mask. If your reactions are similar to mine, you may want to avoid using a full face mask in fast water.

DO'S AND DON'TS!

In any kind of a dredging operation, fast or slow water, **do** become familiar with your surroundings as your first priority. Before you begin work, make sure you know the easiest and most direct route to the surface in the case of an emergency. Don't wait until an emergency happens before you think about this. By then, its too late!

Don't tie yourself into a dredge hole in fast water to keep from being swept down river. It's bad enough having a heavy load of lead attached to your body! If you have to tie something, tie the suction nozzle from a point further up river. Then hold onto the nozzle to keep yourself steady and in place, while you get the hole started.

Generally, the most effective way to maintain your position in fast water is to streamline your body properly, with your head and chest close to the bottom and your rear end slightly elevated. This posture allows the water flow to push you down, toward the bottom, so you can get a better footing. Begin creating your dredge hole as soon as you can. The hole will help anchor you in place. The larger you dredge the hole, the easier it gets.

Some dredgers try to solve their stability problem by putting a lot more lead on their weight belts. Sometimes in turbulent water, more lead can be a help. But, be extra careful when walking out of the water, so you don't overload your ankles and knees and injure yourself.

Most importantly, **don't** solve your fast water buoyancy/stability problem by adding additional weight **belts**. It's hard enough to get one belt off in a hurry, without compounding the emergency with two or three of them! Additional belts tend to shift around so that the quick releases are in different places. Difficulty in finding them can contribute to a panic situation and put your life at risk!

Whatever else you do, early in your dredging career, **do** discipline yourself to **never** try and swim for the surface in an emergency, **while wearing** your heavy weight belt. It just doesn't work! In a panic situation, your body will want to go immediately for the surface instead of removing the weight belt. I have personally saved two people from drowning who were trying to swim for it with their weight belts on. By the time they realized swimming wasn't going to work, they were in too much trouble (panic) to get their own belts off!

This doesn't mean you can't get a good footing on the bottom and jump up to the surface for one quick breath of air. You can do that in an emergency, as long as the water isn't too deep or fast. But, your first priority should always be to get the lead weights off as soon as possible.

Keep in mind that you usually cannot see the quick release buckle on your weight belt while under water. This is because your face mask blocks your vision at that angle. So, it is important to **practice** locating the quick release buckle by feeling for it.

It is also very important to keep your belt from shifting around so that the buckle always remains directly in the front. One of the problems we already noted with more than one belt is that the top one tends to shift around. You may also find that it is better to first remove your work glove before trying to release your buckle.

These are all things you must be able to do, quickly and instinctively, before venturing into fast water. A skydiver would never jump out of an airplane without first receiving a lot of practice and instruction in how to find his rip cord. Similarly, a dredger's life should be just as well protected by having a confident ability to release his weight belt quickly in an emergency.

Some of the weight belts on the market also include a suspender harness. The only ones I recommend are the ones that have a quick release, D ring on one of the suspenders, that permits the shoulder harness to come loose when you release your belt. Otherwise, in an emergency, you may find it difficult to get out of the suspenders, even if the waist belt is released.

Do, always, keep an eye on your diving buddy while dredging in fast water. When we dive with multiple dredgers on an operation, it is standard policy for us all to keep track of each other. If one person needs to leave the hole or go to the surface for some reason, he **always** lets someone know he is leaving. Therefore, when a diver suddenly disappears, we immediately go looking for him. A person in serious trouble under water only has about 30 seconds to get it together. This isn't much time. What good is diving with someone else, for the sake of safely, if you are not paying attention to what is happening with him/her, especially in fast water where there is little margin for error?

If all of this has frightened you, that's good! That means I have accomplished my goal of alerting you to the dangers inherent in fast-water dredging. Being alert to, and fearful of, those dangers is the starting point for making your own preparations and contingency plans for dealing with them - **before** you start working in fast water.

What is fast water? It depends on the individual. An experienced diver might be much safer in a storm of fast, turbulent water than an inexperienced one would be in slow, shallow water near the bank. The key for each person is to begin learning in a safe and comfortable environment, gain valuable experience over time, and never attempt to do anything that you cannot easily handle, with safety.

4

BOULDERS AND WINCHING

It takes an incredible force of water to move boulders in a river. Once they are moving in a flood storm, they can deposit in low-velocity areas, just like gold does. But since boulders do not have a higher specific gravity, mass for mass, than most of the other streambed materials, they can be washed downstream just about anywhere in the river during a major storm. So, you should not use only the presence of boulders to guide you in sampling. You would be much better advised to focus your attention on the boulders that have been deposited along the common path of gold's travel.

In shallow material, you can sometimes detect where important bedrock changes are located, by noting where the boulders have deposited in the waterway. For a boulder, or a group of boulders, to be found in a specific location along the river's path, there may be a sudden bedrock drop-off, a large crevice, or some other kind of lower-velocity condition in that area which caused the boulder(s) to be deposited where it was (see figure 4-1).

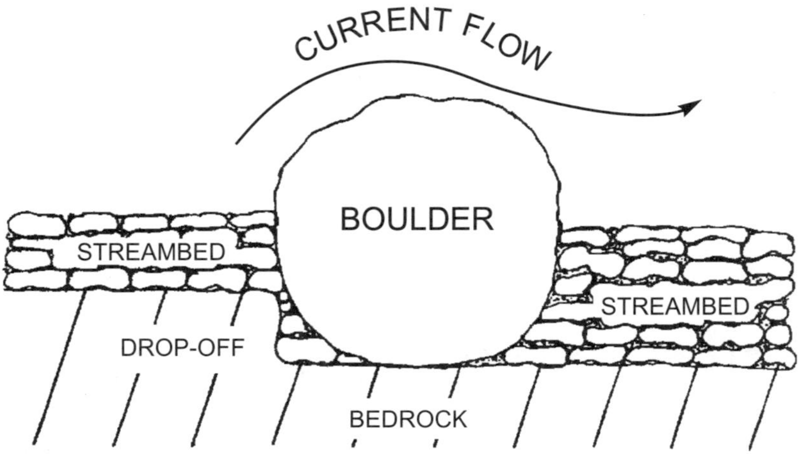

Figure 4-1: You can often find a bedrock change where you see that a boulder is resting in shallow material.

If the boulder(s) is located somewhere along the common gold path, that is a prime spot to do some sampling. Don't limit your sampling to the area just behind (downstream) the boulder(s) though. Go around to the upstream side of it, as well. Look for any bedrock change that may have caused the boulder to stop there. If gold has moved through that area, that same bedrock change could also have caused the gold to deposit there.

You do not always find boulders with every rich pay streak deposit. But, it is not uncommon to find many boulders keeping company with a good pay streak. When you do find them, most of them will probably have to be moved out of your way as you work forward through the deposit.

MOVING BOULDERS BY HAND

There can be a lot of gold deposited under and around the boulders located within a pay streak. To get the gold out from under a boulder, and into your suction nozzle, usually the boulder has to be moved at least a little bit.

One of the most useful tools that can help a dredger move boulders is the five-foot (or longer) steel pry bar. If a boulder is too large to be moved to the rear of the dredge hole by hand, it can sometimes be rotated around to one side, so you can dredge out from under part of it. Then, it can be rotated around the other way to access the remaining gold and material beneath it. A pry bar can be a big help to you in moving boulders around in this way.

The key concern while working around boulders is *SAFETY*! Loose boulders in and around a dredge hole are the gold dredger's greatest danger, especially when working alone. Boulders resting up in the streambed material are often more dangerous than those resting on the bedrock. But, those on the bedrock can cause trouble, too, if they are loose and able to roll - particularly, if the bedrock has any slant to it.

As they are uncovered in a dredge hole, loose boulders should be moved and safely secured, as your **first** priority. They should be placed at the rear of the hole, if possible. But wherever they are placed, they should be positioned so they no longer pose **any** threat of rolling into the hole and on top of someone working there. You can place smaller rocks and cobbles under the boulders as necessary, to make certain they won't roll.

If you start to uncover a boulder that is resting up in the streambed material, **do not** forget about it. Until it has been moved and secured safely, it should be foremost in your mind. If it is not yet ready to be moved, and you still need to dredge around it to free it up, it's good practice to place an arm or a shoulder against the boulder, so you can feel if it starts to loosen up in the material. **Do** not place your arm, head, or shoulder near the **underside** of the boulder, however. Because sometimes a boulder will loosen up and crash down very quickly, without warning. Physical contact with the boulder is helpful when you cannot keep your eyes on it at every moment. The face mask limits your visual perception underwater - especially when you need to watch what is going up the suction nozzle.

The main concern here is to take all necessary precautions to keep yourself from becoming pinned beneath a boulder in your dredge hole. If a boulder pins any part of your body to the bottom, it may be difficult, or impossible, for you to get the necessary leverage to move the boulder enough to get out from under it. And, if you are working by yourself ... ? I know of two dredgers who ended their careers in just this way.

Also beware of fractured bedrock walls that tower over you. They can fall apart and drop in your dredge hole as you remove the streambed material that holds them in place. I got pinned once by a slab of bedrock that broke free of a wall as I dredged away the material that was holding it there. Luckily for me, it landed on my steel-tipped boot, and that I was dredging with another guy on that day!

The second greatest danger to a dredger usually comes from a cobble falling off the side of the dredge hole and hitting the dredger in some way -- like on top of the head. This can finish off a dredger just as surely as a boulder. Or it may cause some serious pain/injury. Fingers, and other body parts, can get smashed if you are not careful!

Most trouble with cobbles and boulders can be prevented simply by taking your hole apart with safety in mind. The proper method of taking a dredge hole apart has already been fully covered in *Basics of Successful Gold Dredging*, so I won't repeat it again here. But, as a point of emphasis, the fastest way to take apart a streambed also happens to be the **safest** way.

Production dredging goes very smoothly and quickly when you have some area of exposed (dredged) bedrock between the non-dredged material in front of you, and the cobbles, boulders, and tailings behind you. When you start uncovering a boulder in the material, you should immediately begin planning where to move it, once it is ready. Then, make preparations for finding room for it in the dredge hole.

If you run across an occasional boulder that you can't move by hand, sometimes you can dredge the material (and gold) out from around and under it, without trying to move it out of the dredge hole. In this way, if there are not too many boulders, you can keep moving forward on the pay streak without them slowing you down much.

But if there are a lot of large boulders down along the bedrock, you may find that they will have to be moved if you are to uncover very much bedrock with your suction nozzle. They will also have to be moved to make room for other boulders as you move forward on the pay streak. If you cannot move the boulders by hand, then you will need some mechanical assistance.

DIFFERENT TYPES OF WINCHES

COME-ALONG: A "come-along" is a portable, hand-operated winching device that can be of considerable help to a small-sized dredging operation. It can be used to move those boulders that are not huge, but which are too large to be moved by hand. Come-alongs range in size. I recommend a larger version of the better quality, rather than the really cheap, imported models. Come-alongs have a great accessibility factor, because you can carry one just about anywhere. Their operation is somewhat slow. But they **will** give you that extra edge when you need to move just a few boulders, and you don't own, or want to set up, a power winch.

GPIP-PULLER: Several companies make a hand-winching device that rides along on a steel cable.

There is a handle that you crank back and forth, similar to a come-along. Each time the handle is cranked, the device moves an inch or two along the cable. These units also come in different sizes. In my opinion, grip-pullers are a good step up from a come-along, both in pulling power and dependability. I have used these units under water, but find that the water resistance adds substantial work to the cranking action. Still, when you are working alone, having this device in the hole with you allows you to see what the boulder is doing while you are winching it

USING YOUR TRUCK AS A WINCH: If your vehicle can be driven to a nearby position, you can stretch a cable from the vehicle to the boulder. The proper direction of pull can be accomplished by running the cable through snatch blocks (heavy-duty pulleys) which can be anchored to trees, boulders, or whatever is available. Then, you can use the pulling power of your truck to help move the boulders out of the way. Four-wheel drive vehicles work better for this, especially when they are carrying a load to increase tire traction.

When conditions are right for it, using a vehicle to pull smaller boulders can be much faster than using a come-along. One person operates the vehicle, while a diver is in the water, slinging the boulders. Safety becomes a greater concern when more than one person is involved in the pulling and the slinging of rocks. Communication and coordination between the "puller" and the "slinger" are very important to prevent serious accidents.

Larger-sized trucks, tractors, bulldozers, and other heavy equipment may be used to move bigger rocks with even better results.

AUXILIARY TRUCK WINCHES: Auxiliary automotive winches are also able to move small to mid-sized boulders, for a dredging operation, with excellent results. Some of those little winching units have a wondrous amount of power. A typical 8,000-lb electric winch will move a surprisingly large boulder!

If you are going to be using an electric winch, you may want to consider installing dual batteries in your vehicle. And, it helps to keep the engine running while you are winching, so the batteries can quickly regain their charge between pulls.

When using a truck-mounted auxiliary winch to pull boulders, it is a good idea to block all four wheels. This precaution helps to prevent the vehicle from moving, instead of the rock you are trying to pull. The front tires should be blocked especially well. If there is an embankment to worry about, it is also a good idea to run a safety chain or cable from the **vehicle** to some additional anchor, like a tree or another truck. This will protect against losing your vehicle over the embankment if the boulder happens to roll the wrong way. Sometimes this added measure is necessary just to keep the vehicle from sliding during the pull.

Probably the best kind of auxiliary vehicle winch for dredging purposes, is one that can be attached mechanically to the "power take-off unit" on your vehicle's transfer case, if it has one. With this kind of winch, you can use the full power of your truck's engine to pull rocks. Not only will this provide you with more pulling power, but you will not have to worry about your batteries running down.

There is also a lot of good to be said for the portable electric and hydraulic winches on today's market. I know of many recreational and commercial dredgers who use them. They are quite powerful! Portable electric winches can be framed up to (1) attach to a vehicle, (2) be taken out to the dredging site, or (3) even be floated on a platform, where there is deeper water. All that is needed is an automotive battery and a means of keeping it charged up. The portable hydraulic winches can be similarly effective. The winch controls can be extended out on a longer cable and even modified to work under water, if necessary.

LOG SKIDDERS: Most gold-dredging country is also logging country. So, there are quite a few log skidders around that can be leased or hired to pull boulders for you. Log skidders are usually equipped with powerful winches. They can really do a job in pulling the larger boulders! Also, you can drive them into some pretty difficult areas. Hopefully, you can work out a deal that will be ideally suited to your operation. One, perhaps, where you pay by the hour and have the skidder arrive at a certain time each day, or whenever you are in need of its services.

PORTABLE POWER WINCHES: Gasoline-powered, mechanical and hydraulic winches are available in all sizes. Generally, the larger they are, the more winching power they produce. But the added size and weight also makes them more difficult to pack into some of the less accessible areas.

If you find a wide-spread, rich, pay streak, with plenty of large boulders that need to be moved, you should set up on your dredging site the most powerful winch you can haul in there. The more pulling power you have available, the smoother and faster winching will go.

I never recommend that a dredger buy a portable power winch just because he will be dredging. It is much better to wait until you know exactly what your needs are. I have worked many pay streaks where no winching was needed at all. And, in many of the pay streaks that did require winching, a number of the boulders were so large that the small, portable, store-bought type of winch would not have been adequate for the job.

You can find some pretty heavy-duty used winches for sale at about the same cost as a lighterweight, new, portable one. I advise waiting to see what you will need before putting your money into a winch.

SETTING UP A HOLE FOR WINCHING

Before you start winching boulders, it's a good idea to first make sure that you have located the lower (downstream) end of the pay streak. You do **not** want to winch boulders onto any part of the pay streak that you will be working later. While the winter storms could possibly move some of the cobbles and dredge tailings which were placed on a pay streak in error, it takes **much** more than a winter storm to move boulders from where you winch them. It's much better to plan ahead, so you only need to move boulders **one** time.

BUILDING A RAMP

It takes a **tremendous** amount of winching power to pull a boulder up and over another boulder from a dredge hole. The direction of pull is all wrong, and the second boulder will act like a barrier, as shown in figure 4-2.

The winching is much easier, smoother, and far less dangerous, if a ramp is built so the boulders will have an inclined runway to be pulled along. Cobbles can be used to make an effective boulder ramp, as shown in figure 4-3.

Therefore, an important part of setting up your dredge hole for winching is to construct a ramp/runway that can be used to easily remove the boulders, without them encountering any further obstructions.

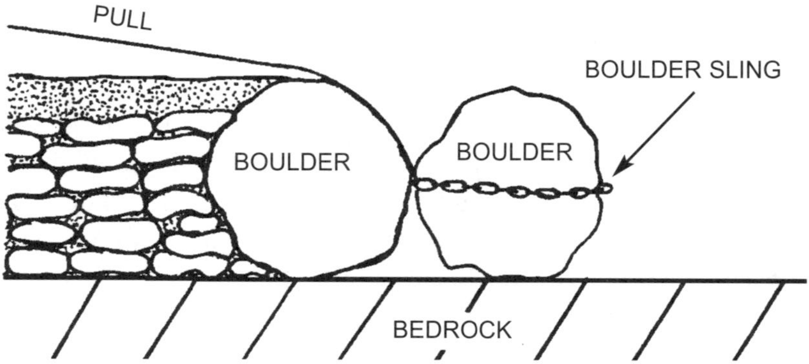

Figure 4-2: It can be really tough to pull a boulder out of a hole without a ramp.

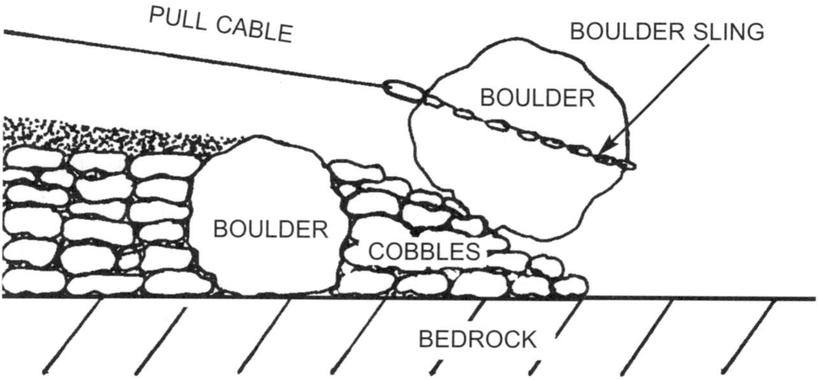

Figure 4-3: Cobbles can be used to make an effective ramp for winching boulders out of a dredge hole.

Another approach to setting up your hole is to dredge around a number of boulders, as much as possible, to free them up for winching. This way, more boulders can be pulled out of the hole once winching is begun. Then you can come back later to clean up the bedrock with your dredge.

To complete this job effectively, you will need plenty of air line attached to your HOOKA system. It will be necessary for the diver to sling each boulder and follow it to its winching destination, where he will disconnect it. Then the sling and cable must be pulled back, to be reattached to the next boulder. Since this process requires a lot of movement, depending on the distance involved, you may need to attach an extension onto your air line to accomplish this smoothly.

FEASIBILITY OF MOVING BOULDERS

A dredging operation that requires a lot of boulders to be winched is going to move slower than if no winching were necessary. A winching operation almost always requires the involvement of at least one other person, sometimes even two or more. Those "extra hands" may expect to get something for their active participation in your dredging project. So you will find that when you need to use a winch, you will be moving more slowly through the pay streak, and it will cost more to run the operation. Therefore, a pay streak that has lots of boulders usually needs to pay pretty well to make the additional effort worthwhile.

Sometimes, you may hit an excellent pay streak, but find that it lies beneath many boulders that have to be moved. Even though there may be plenty of gold under the boulders, after figuring the time and expense involved in winching, you could conclude that the gold recovery, though excellent, is still not sufficient to make the project economically feasible. Big boulders are not easy to move. Moving them takes time and money. More, sometimes, than the gold is worth. So, to avoid getting bogged down in an unworthy project, keep track of your daily expenses and your daily gold recovery. Then you will have some basis for calculating whether it is financially worthwhile to continue dredging in that area.

If your winching helpers are not full-timers on your dredging operation, and if the location is not too remote, you might arrange to have them arrive at a certain time each day to help pull boulders. That way, you can spend the morning setting up the hole to winch as many boulders as possible. Then, when your help arrives, the boulders can all be winched out of the hole at one time. After winching is completed for the day, you can go back down and clean up the hole with your dredge. This way, you are not paying helpers to stand around with nothing to do while you are dredging. It can make a difference.

SETTING UP A WINCH FOR OPERATION

When you are using portable power winches to move boulders in a dredging operation, they **must** be set up on a solid and stable foundation. It takes a tremendous amount of force to move a boulder. Sometimes, the boulder will move quickly. Or, sometimes, the sling will slip off the boulder, causing the cable to suddenly go slack. Maybe the boulder will loosen up and roll the wrong way, causing a sudden, heavy stress on the cable. When these things happen, and they do often, you don't want your winch bouncing around or sliding off the platform. That could be extremely dangerous! The winch **must** be stable.

The winch also should be anchored to a solid object behind it, which will hold its position much more securely than the boulders that are to be moved. A fair-sized tree, or a large boulder, directly behind the winch platform can work well for this purpose.

Also, the cable or chain being used to anchor the winch must be considerably stronger than the winch's ability to pull. The last thing you need is your anchor chain or cable breaking while you are pulling a boulder! This could cause the winch, and the winch operator, to be yanked off the platform, resulting in a serious accident. Undamaged, heavy-duty truck tow chains, with the adjustable end hooks, are excellent for anchoring a winch. Make sure you get chains that are strong enough.

Your winch should also be anchored to a point which is **lower** than it is. If the winch is anchored from a low point, when boulders are being pulled, the winch will be held down more solidly onto the platform. (See part A of figure 4-4) If the winch is anchored to a higher point, during pulling the winch can be lifted off its platform. This, also, can be dangerous (See part B of figure 4-4.).

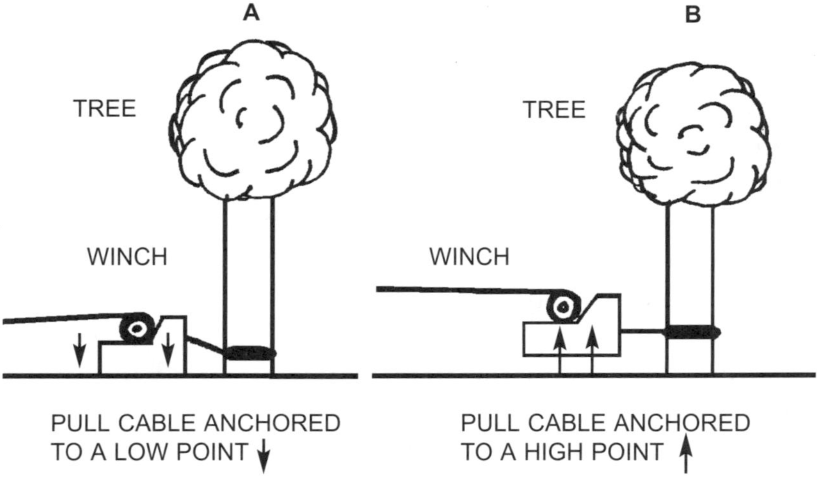

Figure 4-4: The winch needs to be anchored to a low point or it will be lifted off its foundation when boulders are being pulled.

Sometimes you will not be able to find a good location for your winch along the bank. You want a location (1) where the winch can be set up to directly face the boulders to be pulled; and (2) which provides a level, solid foundation, with a properly located, fixed object from which to anchor the winch. When this happens, you can almost always find an acceptable foundation, somewhere between two large objects along the bank. The platform may require a little concrete work to get it right. The winch can be anchored to one of the objects, while the other object can be used to attach a snatch block (heavy-duty pulley). The cable can then be run from the winch, through the snatch block, to the boulders that need to be moved. (See figure 4-5).

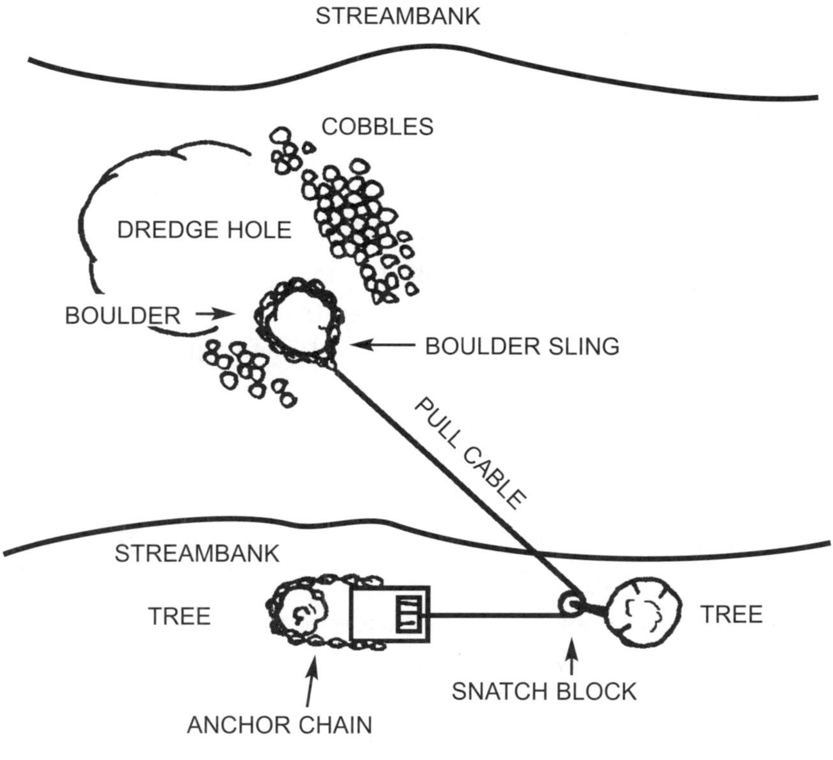

Figure 4-5: Positioning winch between two anchors on the bank because there is no easier way to get a straight pull with the winch.

In this situation, the snatch block should also be anchored to a point lower than the winch (for the same reason that the winch anchor needs to be attached to a lower point).

One advantage to this rigging setup is that if the cable does happen to break under a great strain, it will be less likely to fly directly at the winch and its operators.

When you set up a winch, position it so the boulders will be pulled, as much as possible, in the desired direction. The boulders will also need to be pulled far enough away from the dredge hole that they will not have to be moved again at a later time. Often, you will not be able to set up the winch in a position where you can directly pull the boulders in the desired direction. For example, you may want to pull them downriver, rather than toward the stream bank. This situation is most commonly corrected by using directional-change snatch blocks. A snatch block (pulley) can be anchored at a point from the exact direction that you want to pull boulders. The cable then runs from the winch, through the snatch block, to the boulder. In this way, you can usually move the boulders in any direction you desire, from a stable winching position along the bank, as shown in figure 4-6.

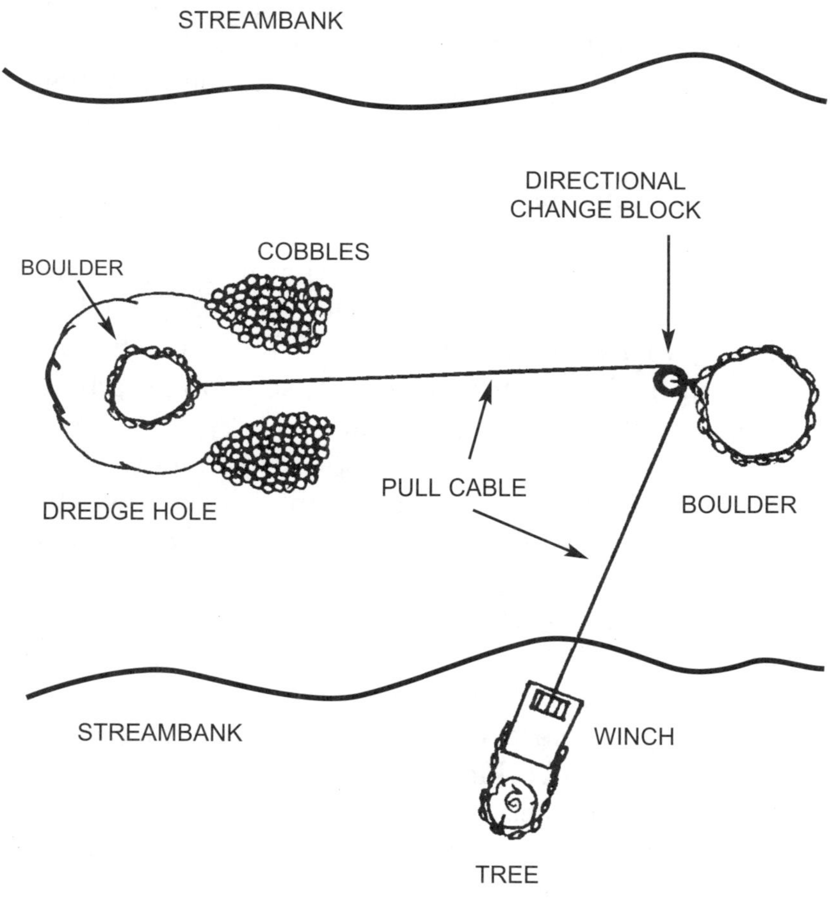

Figure 4-6: Setting up the proper direction of pull by anchoring a snatch block directly downstream in the river.

Directional-change blocks for winching should be very heavy duty. They need to be stronger than the cable being used, or the ability of the winch to pull.

The best pulley is one that can be quick-released from the cable. This way, you don't have to feed all the cable through it to get rigged for winching. Good winching pulleys are generally available at industrial equipment supply stores and at marine equipment shops.

Directional-change blocks can be attached to trees or boulders with the use of additional cable or heavy-duty chain. You will find that a few extra tow chains (long ones) will come in very handy while setting up a winching operation. Chains which have the end hooks on them are best, so you can adjust their length to meet your needs.

To minimize damage to the environment, place pieces of wood between the cable/chain and the trunk of any tree that you use as an anchor. When you anchor a directional-change block to a tree or a boulder, you will be better off anchoring to a low point. This will reduce the chances of rolling the boulder over or pulling down a tree.

When you use a cable to anchor a winch or a directional-change block to a boulder, set it up so that the pulling action will not cause the cable to be pinched (between two boulders, or between a boulder and the bedrock) or be pulled into the material underneath the boulder. Otherwise, after the winching, you may have trouble retrieving your cable without having to move the anchoring boulder as well. (See figure 4-7.)

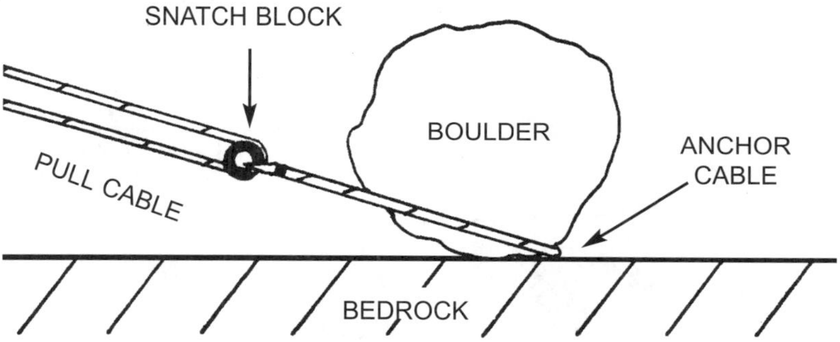

Figure 4-7: Avoid pinching an anchor cable between things or you'll have trouble getting it back again after you're through winching.

At times, you may need to set up a directional-change block, to pull boulders in a desired direction, but cannot find a downstream boulder that's large enough to use as an anchor. In this case, you can run an additional cable from a fixed object further downstream, over to the opposite side of the river, and attach your directional-change block to the cable. Then, by increasing or decreasing the length of the cable, you can position the block right where you want it, as shown in figure 4-8.

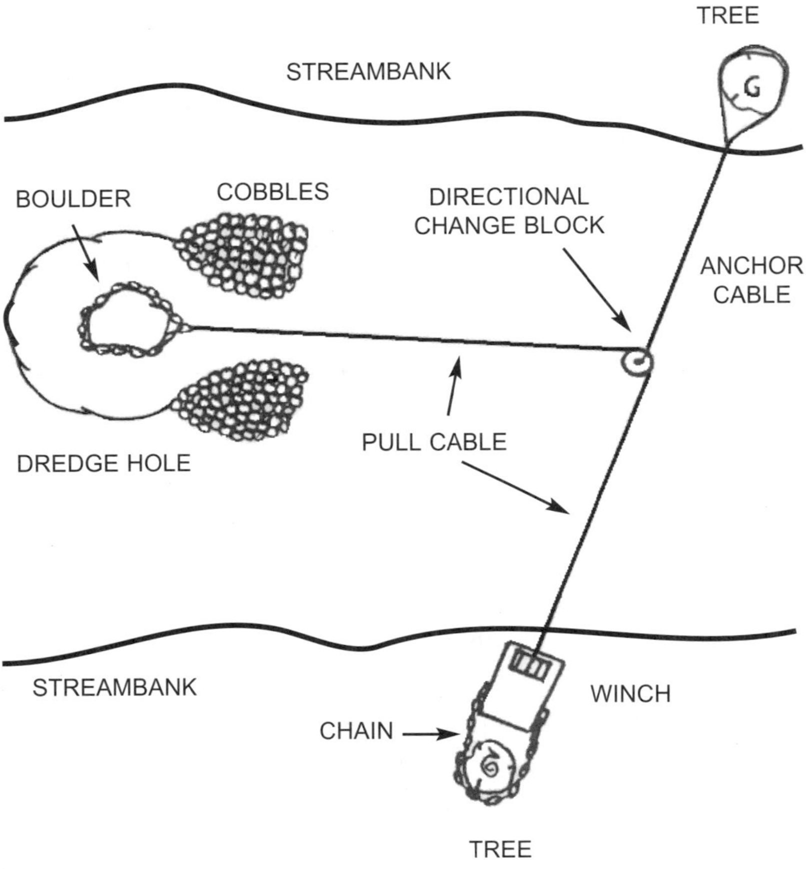

Figure 4-8: Setting up a directional-change block by extending it out on a cable from the other side of the river.

It's usually a bad idea to winch boulders with rope, even the smaller ones being pulled by your truck or a come-along. When you use rope for winching, even if it is doubled multiple times for extra strength, the rope stretches. And it stretches. And it stretches. Then, it breaks. Such breaks can be dangerous when working around boulders. Also, it can quickly exhaust your supply of rope. Rope is just not strong enough for the job. Steel cable is best. You might get a deal on good, used, steel cable from the scrap metal yards. You'll have to call around to find out who has it when you need it. Scrap yards usually sell it at scrap prices, by the pound, even if it is in good condition.

The cable you use on your winch should **always** be considerably stronger than your winch's pulling capacity. The last thing anyone wants to see is a bunch of out-of-control, steel cable flying wildly back in his direction. Faulty/worn cable should be replaced immediately and never used thereafter for winching.

Winching boulders involves an incredible amount of force. So does hauling logs by cable. Similar cables are used in logging operations to pull trees to the loading area. I've heard stories of logging cables snapping, flying back, and cutting a man in two. You are dealing with a similar amount of force when winching the larger boulders in a dredging operation. To be safe, be sure to use only cable that is stronger than your winch's capability of pulling. Then you won't have to worry about it. (At least, not as much.)

WINCHING SIGNALS

During a winching operation, it will be necessary to have a diver in the water to set the sling on the boulders. The diver will also need to remove the sling from each boulder, after it has been moved, and return the sling and cable back to the dredge hole for use in moving the next boulder. This process will continue until all the boulders, for that stage of the dredging operation, have been moved.

Unless the diver is slinging boulders **and** operating the winch (a very slow way to go, unless the controls to the winch are in the water with the diver), another person must operate the winch, truck, or other device that will be doing the pulling. Once you have more than one person involved in the operation, communication becomes a **critical** part of the process.

If the winch is a smaller one (or if a truck is being used to pull boulders), and the pulling position is in sight of the "target" boulders, the job might be accomplished with only one winch operator. On the other hand, in a normal "two-man" operation, if the winch operator is unable to keep fairly constant eye contact on the area where the winching is taking place, perhaps a third person will be needed to help with the communication. Each situation will be different. Since this need for accurate communication is a safety matter which requires on-site judgment, you will have to decide, in each case, how many people are needed to winch safely.

Sometimes a truck can be turned around to pull

backwards, and the driver can directly see, and simultaneously carry out, the diver's signals. And the controls on an electric winch will often allow a second person to be positioned well enough to see what is happening where the rocks are being moved.

While a boulder is being moved, the diver can watch its progress and signal the winch operator to stop pulling (and/or give slack on the line), so the boulder sling can be adjusted, if necessary, to successfully complete the movement as planned. The main point is: the diver must communicate to the winch operator quickly and without error. The greater the pulling power and/or speed of the winch, the more important it is that these signals be accurately received and acted upon quickly. Imagine that you are pulling a large rock, using a powerful winch, with a heavy steel cable and solid anchoring objects. If your boulder gets jammed so that it will not move, and you continue pulling with great force, **something** is eventually going to give. And, it might not be the boulder! This uncertainty is what you want to avoid.

If you are using a winch that does not automatically feed the cable evenly onto the drum, the operator will sometimes need to manually guide the cable. Otherwise, on hard pulls, if the cable starts crossing itself and is allowed to pinch itself on the drum, it may become damaged and, thereafter, be dangerous for further winching. For this reason, the winch operator may need to focus a lot of attention on the winch, rather than on the diver. In this type of situation, a third person is definitely needed. Someone needs to be watching for the diver's signals at all times. A third person can be at the winch, and can relay the diver's signals to the winch operator.

Gasoline-powered winches make noise. So does the gasoline-powered HOOKA air system supplying air to the diver. The diver also generally has a regulator in his mouth. With all of this noise present, and the diver having his mouth full, verbal signals are usually not very dependable - especially, if there is a substantial distance between the diver and the winch operator. For these reasons, I have found visual signals to be more trustworthy, particularly when there is a person stationed at the winch to relay the diver's signals to the winch operator.

In shallow water, hand signals can usually work pretty well. You really only need three of them: *"PULL; " "STOP; "* and *"GIVE SLACK ON THE CABLE. "* Design your signals so that: (1) they can be clearly understood quickly; and (2) one signal cannot be confused with another. There is a section on winching and signals in my video, *"Commercial Gold Dredging."* You might want to check it out to get some ideas.

At times, you may be faced with the need to winch boulders out of a dredge hole located in deeper water. In this case, the diver will be heavily weighted down to stay on the river bottom. He may also be some distance from the stream bank. In this setting, it may be nearly impossible for the diver to surface to give timely signals to the winch operator.

The process can simply take too long! When this is your situation (and if the water is not moving too fast), you may consider using a buoy, tied to a rope that is anchored near the dredge hole, to relay your signals. I have found the best and safest signals to be: (1) buoy underwater means, *"PULL;"* (2) buoy floating at the surface means, *"STOP PULLING;"* and (3) buoy bobbing up and down in the water means, *"GIVE SLACK. "*

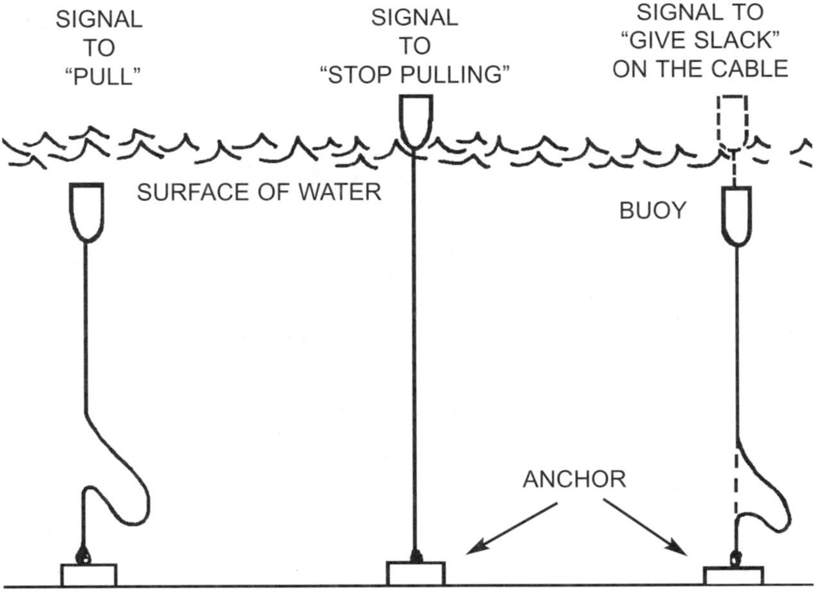

Figure 4-9: A buoy can be used for winching signals in deeper water.

This method of buoy signals is relatively safe. If the buoy is anchored a short distance from the boulder, in order to pull the buoy underwater and hold it there, the diver would have to be **away** from the boulder on the *"PULL"* signal. If the buoy is floating, the diver can be **anywhere**, which is why it is the best choice for the *"STOP"* signal. You need to make certain, however, that nothing is allowed to snag the buoy's rope which could pull it underwater and cause a false *"PULL"* signal!

One other safety note: All divers must **always** watch their own HOOKA air lines during winching, to make sure the line is not snagged up in the cable or rolled over by the boulder as it is moved.

BOULDER SLINGS

When slinging boulders, make sure the pull cable will not rub or bang against other boulders. It's always best to protect the pull cable and let the boulder sling take the pounding. It's going to be pounded anyway. Because of this rough duty, boulder slings should be replaced periodically.

One type of boulder sling often used out in the field is a long, heavy-duty tow chain, with end hooks which allow the chain to be quickly and easily adjusted to any length. This system gives you a fast set up. Just wrap the chain around the boulder in the proper place, connect the end hook to give you the right fit, and she's ready to go. (See figure 4-10.)

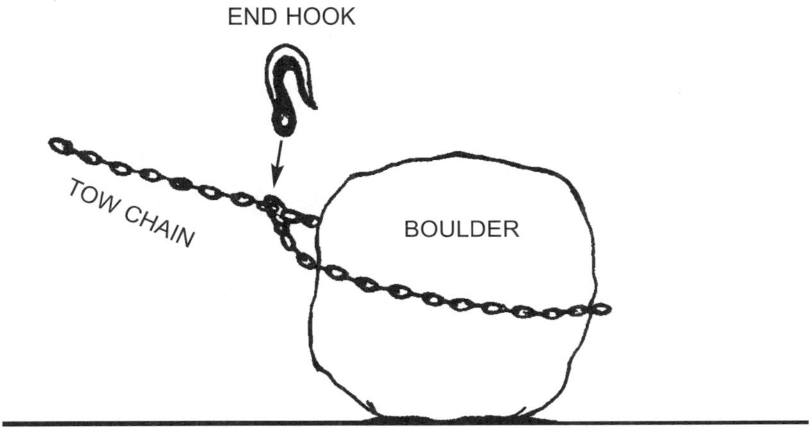

Figure 4-10: Tow chain boulder sling with end hook.

However, some boulders are smoother and rounder, which makes it more difficult to get a good "bite" with a sling made out of chain. Every time you start to pull, the boulder might move just a little and then the chain slips off. You can spend a lot of time working on a single boulder in this way, and it can get quite frustrating.

Logging cable-chokers are useful for making a cable sling that will tighten up on the boulder as you pull. This may improve the situation, but still not work problem-free on round, smooth rocks.

I have found the best remedy, for the round and smooth boulders, to be an auxiliary "boulder harness". This homemade boulder harness is shown in figure 4-11. It consists of heavy cable, chain, steel pipe, and cable clamps. And it is very easy to construct. The sections of steel pipe slide onto the cable, to protect it and to keep the cross chains properly positioned. The harness is set up like a lasso. It pulls tighter around the boulder as stress increases on the line. And, it generally works well in pulling even the most difficult boulders.

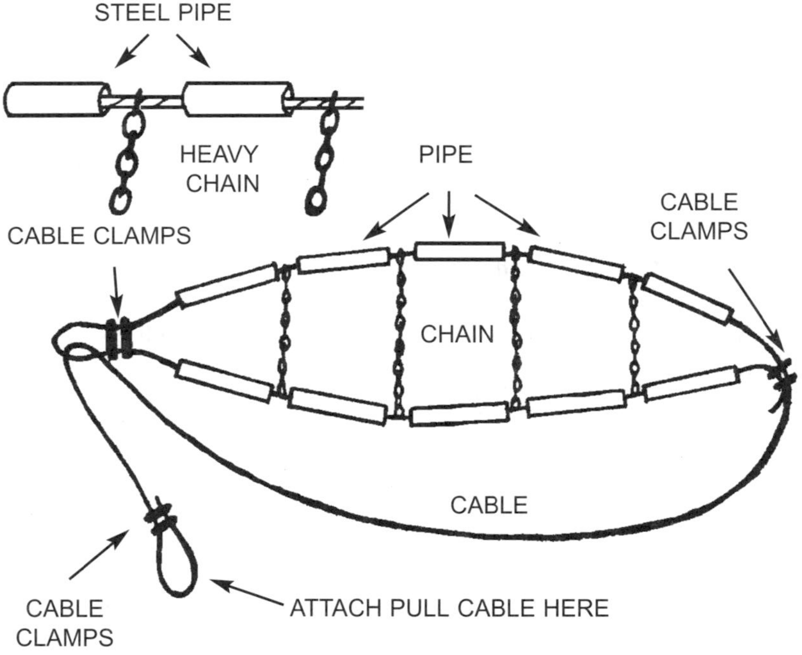

Figure 4-11: How to put together an excellent boulder harness for winching.

PULLING BOULDERS

Some boulders come easy and some do not. A lot of the problem is in breaking the boulder's suction/compaction in the streambed. If your winch does not have the power to pull a certain boulder the way you have slung it, sometimes you can break the boulder free by using a rolling hitch A "rolling hitch" is rigged by slinging the boulder backwards, then running the chain or cable over the top of the boulder, as shown in figure 4-12. This places the winch's pulling power along the most-leveraged position on the boulder, and it will sometimes free the boulder by rolling it.

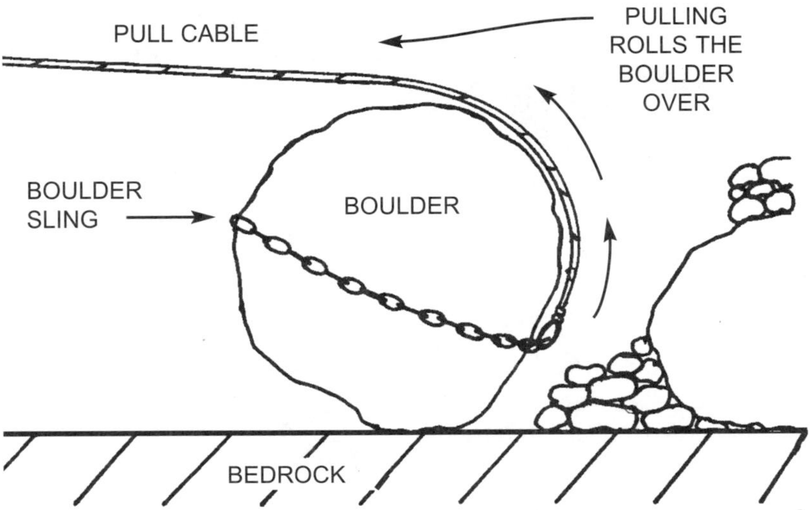

Figure: 4-12: How to sling a rolling hitch.

When pulling boulders up and out of a dredge hole, you should pull them some distance away from the hole. Otherwise, if there are more boulders to be moved, they may begin backing up along your ramp and block the passage of any more boulders. This could require you to move them all again, which is a time-waster that you can avoid with proper planning. (See figure 4-13.)

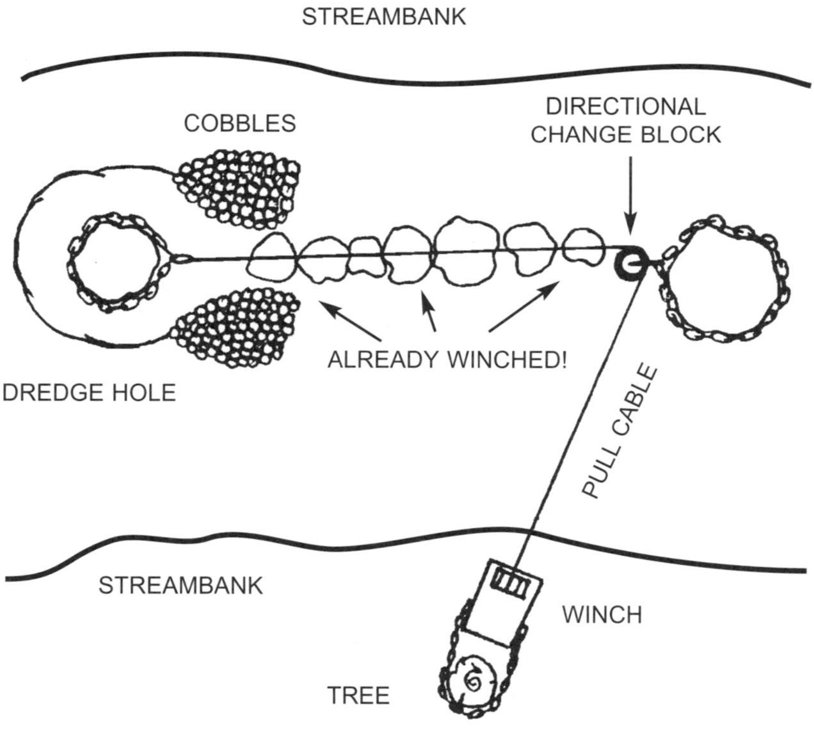

Figure 4-13: If you do not pull boulders far enough away from your hole, they will back up and block further winching.

If you are dealing with relatively deep material and a lot of boulders, you may want to set up an adjustable, directional-change block behind the hole. This way, the boulders can be winched out of the hole in several different directions, which may prevent them from backing up so quickly (See figure 4-14).

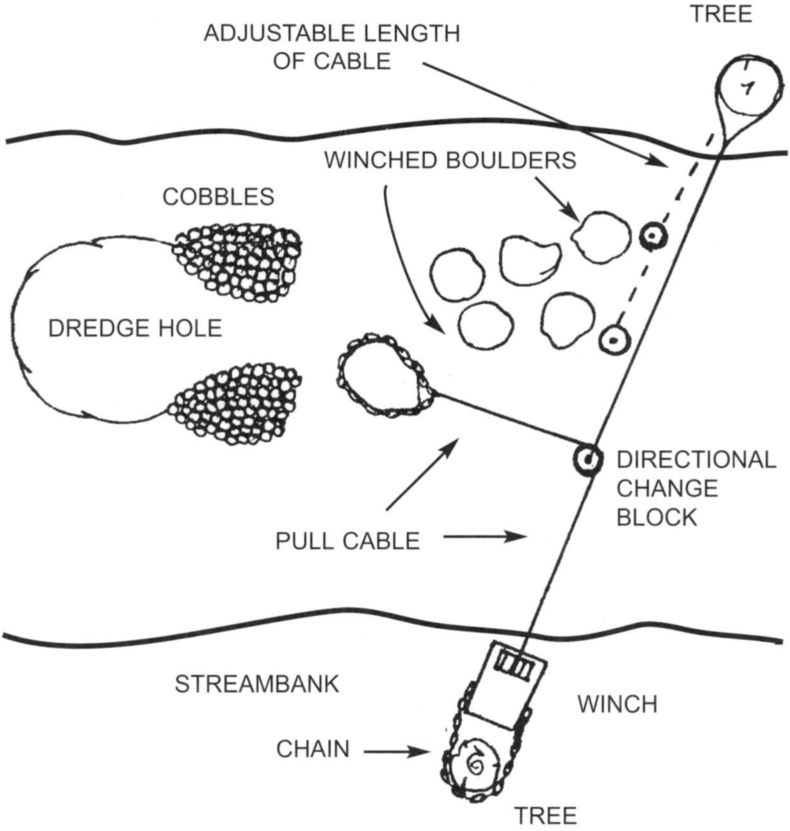

Figure 4-14: Setting up an adjustable directional-change block to pull boulders in several different directions.

Once the dredge hole has been opened enough, some of the boulders can be winched to the back side of the hole instead of taking them immediately up the ramp. Winching will start to go faster when you get to this stage. Depending on the situation, it may be necessary to winch some of the boulders up the ramp and out of the hole, to prevent too much jamming. Be sure to keep access to the ramp free and clear. Otherwise, you may get closed in with too many boulders in the rear of your hole. The more boulders that are jammed up, the more difficult it can be to clear them out of the hole. Even then, they can usually be removed by setting up a directional-change block. This will allow you to move them in a different direction. But since this modification takes extra time and effort, it is far better to plan your hole with some forethought and dispose of the boulders with a minimum of effort at the proper time. You really need to use your ingenuity in planning for continuous and steady progress through a boulder pack. But some pay streaks will pay more than enough to make all the planning and effort worthwhile.

Sometimes, when you have winched your boulder to its destination, it will be resting on top of, and pinning, your sling. If you are using a tow chain as your sling, you can usually just unhook it and have the winch pull it out from underneath the boulder. But if you are using a sling made of cable, you may not be able to pull it out from under the rock without damaging the harness. For this reason, it is a good idea to have a second sling on hand, to help move the boulder off the other harness.

DIVER'S SAFETY

A diver will be safest by staying well away from (1) the area where a boulder is being winched and (2) the path it will be taking as it is being pulled. The forces involved in winching are more than enough to cause a very serious accident. Since the diver is underwater, the winch operator sometimes cannot see what is happening at the site of the winching.

What often happens down there, though, is when the diver sees the boulder getting hung up on things as it is being pulled, he wants to help it along with his pry bar. The less power that your winch provides, the more the diver feels the need to help the boulders in this way. But, it is **not** a good idea to be around a boulder, or be trying to pry it, while it is being pulled! There is simply no good reason to put your life at risk by getting too closely involved with so much pulling power that is not really under your control. The safer courses of action would be to: (1) stop pulling and reset the harness; or (2) reset the direction of pull; or (3) groom/improve the boulder ramp, as needed; or (4) get a stronger winch for the job. Alternatively, you can increase the pulling power of your existing winch by means of double blocking.

DOUBLE BLOCKING

"Double blocking" is accomplished by attaching a snatch block to the boulder sling, running the pull cable through that block, and then back to the last directional-change anchor, as shown in figure 4-15. This type of rigging will nearly double the amount of pulling force that can be exerted against the boulder by the winch and pull cable (Figure 4-15).

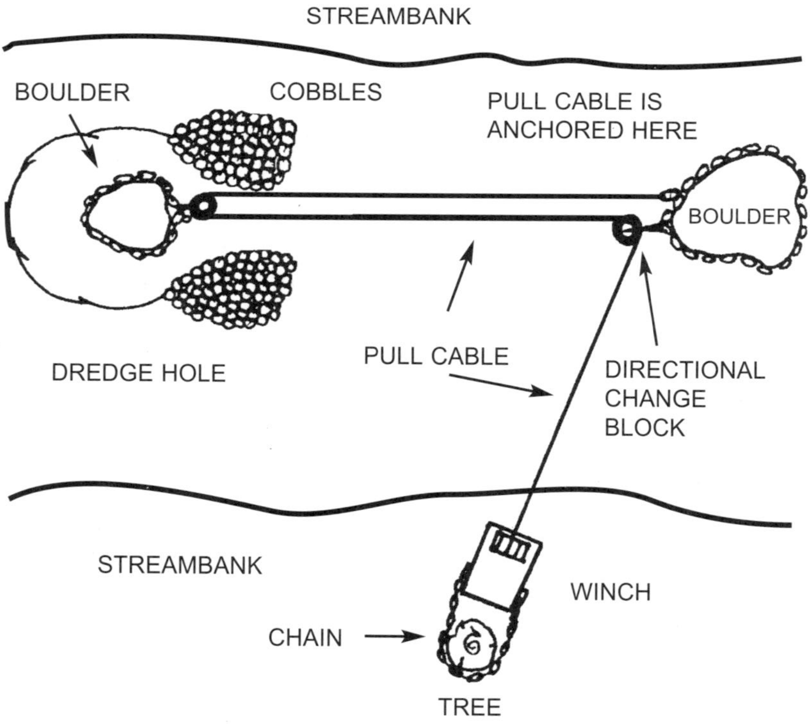

Figure 4-15: Double blocking back to the last directional-change anchor will double the winch's pulling power against the boulder.

If more power is needed, another block can be set up on the line to run the pull cable back to the boulder sling. The pulling force of any winching device can be increased by continuing to double block in this way. There is a disadvantage to all this rigging, however. It takes much more cable to pull boulders any significant distance. Also, it is correspondingly more difficult for the diver to pull the boulder sling and cable back to the dredge hole after each boulder has been moved. With all that cable going back and forth, it can get pretty complicated - especially, if you are dealing with a limited amount of visibility. Quick-release snatch blocks are a **must** when double blocking. This way, you can detach the pulleys from the cable without having to feed it all the way through.

Actually, one double block is not that hard to handle, as long as you have enough cable. It is when you double block a second time that it starts getting difficult. But that alternative is available to you if you need the extra pulling power to move a particularly large boulder.

Directional-change blocks, by themselves, do not give you an increase in pulling power. For a power increase, the cable must be doubled back so that the boulder is moved only half the distance that the winch is pulling on the cable.

BOULDER BOUND

Occasionally, you may dredge down to find a layer of boulders that are so tightly packed together that you just can't get a boulder sling around any of them. And, even if you could sling one of them, the other boulders would probably be in the way to prevent a single boulder from being pulled out to get a hole started. This is called a "boulder-bound hole."

Sometimes you can spread out the hole and find an opening. Your hope would be to just get a few rocks out of there. Then you could build a ramp and haul the others out more easily. But sometimes there just isn't an opening to be found. And the streambed material may be too deep to spread the hole very much. What to do?

A quarry worker from the east coast came up with some good ideas on how to open up a boulder-bound hole. First, you drill a one-inch diameter hole, three inches deep, into the first boulder that you want to move. This is not as hard to do as it might seem. It can be done in about ten minutes, by attaching a rock drill onto a "bedrock slide hammer" and using that as your drilling device. Slide hammers are generally available on the market. Or, they can be made without too much difficulty. You'll want a big one for this job. A rock drill is similar to a chisel, only it is tapered both ways to make a square point (see figure 4-16). It is properly called a "Stone Cutter's Drill". To get the best results, the drill is rotated as the hole is being made.

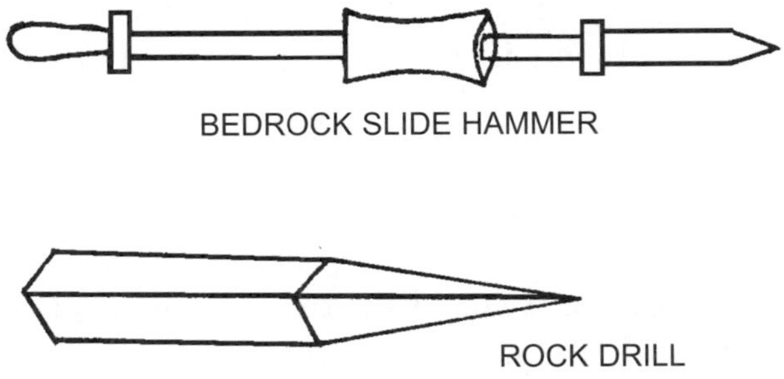

Figure 4-16: Bedrock Slide Hammer and Rock Drill can be used to drill a hole into a boulder underwater.

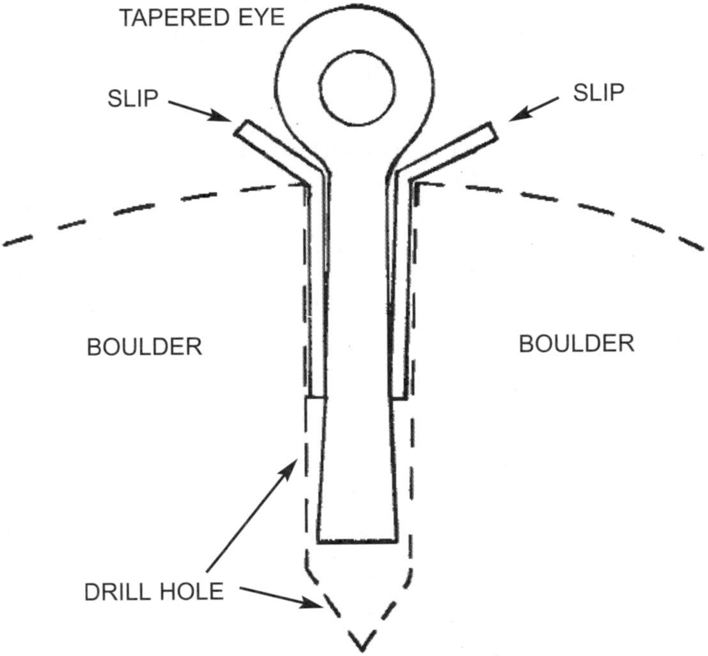

Figure 4-17: Because of the metal slips, when the tapered eye is pulled against, it will anchor more securely into the drill hole in the boulder.

Once the hole is drilled into the boulder, a large, tapered, hardened-steel eye-bolt and two steel "slips" are placed into the hole, as shown in figure 4-17.

The taper on the eye-bolt should be a mild taper. This can be accomplished by bringing a large steel eye-bolt into a machine shop and having them turn it on a lathe for you. Because of the taper on the eye-bolt, and the steel slips placed on either side in the drill hole, the harder the eye is pulled, the tighter it will become fixed into the boulder.

Once the boulder has been removed from the dredge hole, the eye can be removed from the boulder by pounding it back into the drill hole, removing the slips, and then removing the eye.

To get that first boulder out of the hole, you will likely need to pull it upward and out at the same time. This can be accomplished by going to the scrap metal yard, picking up some heavy, steel T or I-beam, and having it welded into a tripod frame. The tripod doesn't have to be very large. It needs to stand slightly taller than the surrounding boulders, to allow you to create an upward pull.

To move the boulder, you set up the tripod with a snatch block, just over the top of the boulder, as shown in figure 4-18. In this way, you can pull the boulder up and out of the hole at the same time. Once a few boulders have been removed in this way, a ramp can be constructed, and the remaining boulders will come out of the hole more quickly.

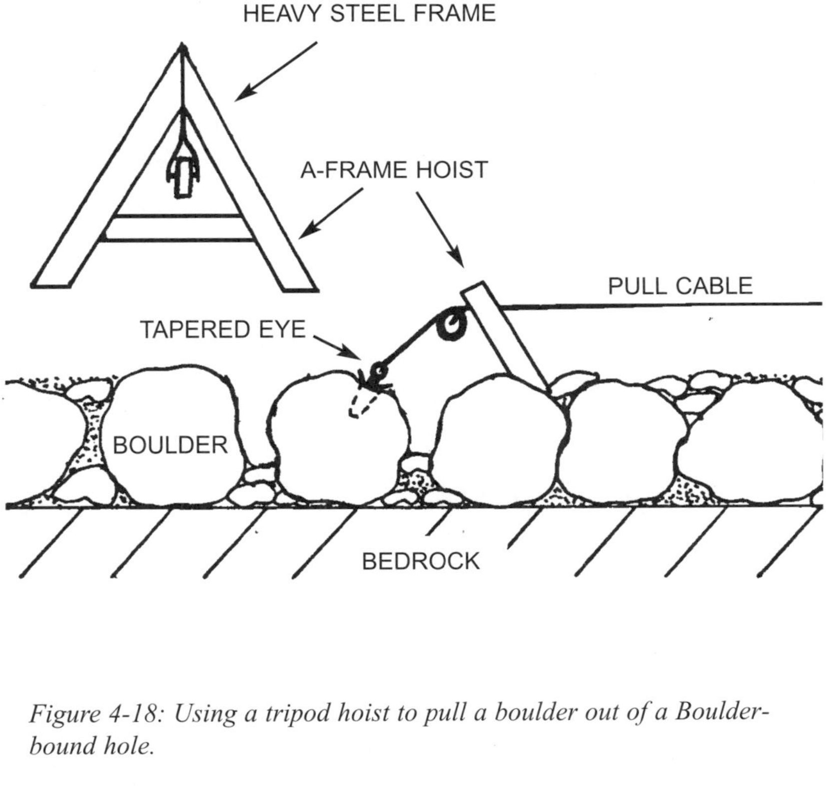

Figure 4-18: Using a tripod hoist to pull a boulder out of a Boulder-bound hole.

5

COLD WATER DIVING

STAYING WARM

It's not at all uncommon for a gold dredger to have to deal with cold water. A dredger might locate a large, well-paying deposit that continues to produce very well as he extends his dredging operation into the fall and winter months. Or, a dredger might decide to go dredging in a location where the water is cold year round, like in the high country or most places in Canada and Alaska.

Actually, because nearly all of the west's gold-bearing areas have a definite winter season, dredging there usually ends, for most folks, when the cold weather sets in. Even so, many gold dredgers will be exposed to at least some cold water, either at the very beginning or very end of the dredging season. Or, for you "animals" out there, during the winter season as well.

Dredging in cold water is more difficult than dredging in mildly cool water. But, with the proper equipment, the difficulties can be reduced to the point where a determined, hardy dredger can make a go of it.

The most important objective when working in cold water is to keep your body comfortably warm. Today, that is mainly accomplished with: (1) wet suits; (2) dry suits; and/or (3) hot-water suits.

Some gold dredgers work through the winter months as well!
Photo by Ron Wendt

WET SUITS

Wet suits are designed to allow water to get inside the suit. Your body heat then warms the water, which insulates you from the colder water outside the suit. In really cold water, the main problem with a wet suit is the breath-sucking, frigid shock you experience as the cold water rushes into your suit when you first enter the water. You also have to relive this unpleasantness every time you reenter the water--after knocking out a plug up in the jet tube or each time you return from taking a break. This "cold-water shock" has a cumulative effect on the body. Even the toughest dredgers may find themselves going "one, two, three," as they "psych" themselves up to reenter the water. They struggle to subdue their memories as they attempt to force their bodies to jump back into the icy water.

Some wet-suit divers ease the pain of cold-water shock with a tub of hot water on the bank. They pour hot water into their suits just before reentering, which helps the transition considerably. Of course, warming systems that provide a steady flow of hot water into a wet suit are much better. But more on that later.

There are several different types of wet suits, some which are designed especially for cold-water use. Cold-water wet suits are generally made of thicker rubber, have few or no zippers, and almost always have the hood attached directly to the top of the wet suit. There is also a "shortie vest," which is designed like a wet-suit version of a pullover, sleeveless T shirt with a hood attached. A shortie vest can be worn beneath a regular wet suit to give a great deal of added warmth. In addition to the extra thickness of rubber around your upper torso, the shortie eliminates the shock of having cold water run down your neck and back. This helps!

For the most part, wet suits are intended to provide protection against only mildly cold water. They are great for this. Extremely cold water is managed much better with the dry suit and hot-water options.

DRY SUITS

Unlike wet suits, which are designed to allow water into the suit, dry suits have special waterproof zippers and seals to keep all the water out. For this extra protection, however, you must pay more. If you buy a dry suit for dredging in cold water, I strongly urge you to buy a GOOD one. Yes, the better suits cost more. But, they are warmer, last longer, and will generally serve you much better as you dredge for gold. With a good dry suit to keep you warm and almost completely dry, you can have fun dredging in cold weather, because you will be comfortable, not cold and miserable.

There are different makes and models of dry suits available on the market. Since dredging requires a lot of arm and shoulder movement (e.g., for tossing cobbles), I recommend you get one that doesn't fit too tight. Try to find a dry suit that has the zipper somewhere other than across the back of your shoulders, because it will be less restricting on your arm movements. And, if you drink coffee, it's a good idea to have a special "relief zipper," so you don't have to get completely out of the suit every time you need to go behind a rock.

Even with a completely sealed dry suit, it will still get damp inside while dredging because the work will make you sweat.

If you have a dry suit and still get cold, a warm liner or clothing worn inside the suit can be a big help. Allowing room for this extra layer is another reason for buying a dry suit that is loose-fitting. Wool is probably best for the liner, because wool continues to "work" when it is wet. Reprocessed wool is good, but 100% virgin wool is even better. When you are shopping for a liner, if it doesn't say "100% virgin wool" on the label, then it's probably reprocessed wool.

Another way to keep the feet warm is to wear battery-operated, electric socks. In really cold water, for those of you with cold feet problems, mild-current, electric-heated socks just might be the solution. I would not recommend these, however, unless your suit is thoroughly dry inside the booties. Electric socks can generally be found at the various department stores and sports shops.

DRY SUIT CARE

The waterproof zippers are the most important and expensive part of a dry suit. You need to be very careful with them. They should not be carelessly zipped up over any protruding obstruction. And, they should be kept out of the dirt and sand.

One key to prolonging the life of a dry suit zipper is to have someone else zip it up for you--especially, if you cannot see it because it is awkwardly located (e.g., along your back or across your shoulders).

Dry suit zippers should also be lubricated regularly. Lubricants which may chemically attack a rubber product should not be used. Lubricants which will adhere to and hold sand, dirt, or grit should also be avoided. Most of the dry suit manufacturers recommend that their zippers be lubricated by rubbing them with bee's wax. Bee's wax can usually be purchased in sewing supply shops.

Sometimes, after extended use, nylon fibers will start to pull out of a dry suit zipper's rubber seals. Over time, they will continue pulling out, further and further, and can become a problem if not attended to. The best way I have found to secure these loose fibers is to melt them back to the rubber by using the flame of a match or a cigarette lighter. You must be very careful, however, not to allow your heat source to get close enough to damage the rubber.

When your dry suit is not going to be used for a while, store it with the zipper completely unzipped. This way, the compression seal will not lose its sealing efficiency by being closed up for a long period.

If a dry suit zipper wears out, or becomes damaged when you try to zip it over an obstruction, replacement zippers are available. You can replace the zipper yourself or send the dry suit back to the manufacturer and pay them to do it.

The neck, wrist, and ankle seals of your dry suit should also be regularly treated with a silicone spray lubricant. Not only will the life of the seals be prolonged, but they will perform better if the silicone is applied immediately before each use of the suit. Spray silicone can be found at most hardware and automotive supply stores.

DRY SUIT MODIFICATIONS

The useful life of a dry suit can be extended by adding protective patches over those areas which will be exposed to more wear and tear during dredging. The knees are a must, since a dredger spends a lot of time moving around on his or her knees at the bottom of the stream. Generally, a dry suit will come with a set of rubber pads on the knees. These may offer some protection for your knees against impact, but they usually are not substantial enough to prevent your suit from being torn, or worn through, from the constant rubbing against the rocks of a stream bed. To protect against this, you should glue an additional patch over each knee. I have found that tire inner tube rubber works very well to protect the knees on wet and dry suits. This rubber withstands abrasion much better than the neoprene rubber that the dry suit is usually made of. It also eliminates the need to wear additional knee pads.

I have trouble with strap-on knee pads. They usually slip around and need constant adjustment. If I tighten them enough so they won't move around, they chafe the backs of my legs. Instead, I like to just glue the extra protection onto my suit and forget about it.

The usual neoprene rubber repair glue that you find in the dive shops does not work well on inner tube rubber. Oil-based, contact cement does much better for this.

An additional modification would be helpful for dry suits having a zipper along the front of the suit. By gluing a protective rubber flap onto the suit, which covers the zipper, you can protect the zipper from being damaged by the heavy contact with boulders that occurs when you are wrestling them around to the rear of your dredge hole. Zippers that extend through the crotch and up the back of the suit can also be damaged by sitting on rough and irregular rocks, both in and out of the stream bed. A protective rubber flap can help out there as well.

This modification can be made by cutting a strip of neoprene rubber that is about three inches wide and long enough to extend over the entire length of the exposed zipper. The flap can be glued to the suit on one side of the zipper and should be positioned so that it will not interfere in any way with the zipper's movement. While installing this flap, be very careful to avoid dropping any glue onto the zipper. Small strips of Velcro can be sewed or glued onto the other side of the flap and onto the suit. These will hold the flap closed and protect the zipper while the suit is being used (See figure 5-1).

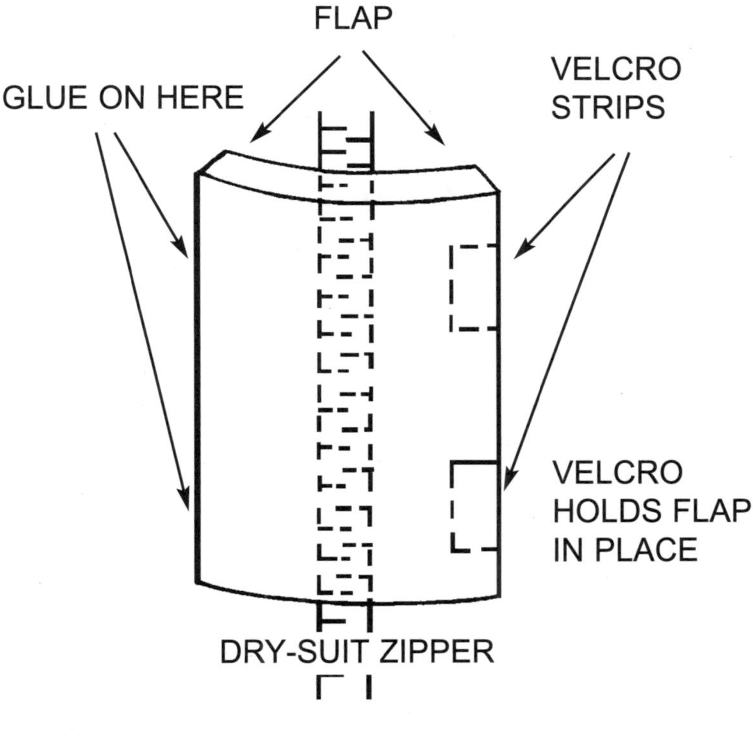

Figure 5-1: Attaching a flap to cover a dry suit zipper where heavy rubbing or sitting on the rocks could otherwise damage it.

CAUTION: Be aware that these modifications to a new dry suit may void the manufacturer's warranty. Some warranties require that no modifications or repairs to the suit be made by the owner. While such modifications can help to protect your suit from the effects of the heavy work encountered in gold dredging, you should be mindful of the impact of such modifications on your warranty. The warranty card will tell you what is allowed. If you still have questions, you should contact the manufacturer.

SUIT REPAIRS

Gold dredging can be tough on wet or dry suits. So it's a good idea to have on hand a supply of repair glue and neoprene patch material when you are dredging in cold water. This way, you will be able to quickly patch leaks as they develop and keep the cold water out of your suit.

There are two kinds of patches which can be used to repair a suit: hard patches (tire inner tube rubber); and soft patches (neoprene rubber). Hard patches are best suited for areas of heavy wear and tear, like the knees. Soft patches are more appropriate for places where extra stretch is needed, like the underarms.

Neoprene-rubber patch material is usually available anywhere that neoprene suits are sold, repaired, or manufactured. The kind of neoprene with a smooth rubber surface on one side and a nylon backing on the other, works very well as patch material. The rubber side will glue and stick well to the surface of the suit. The nylon side will give the patch a protective surface. If this type of rubber is not available, other types of neoprene rubber can also work alright for patching. You may have to take what you can get.

If appearances are important to you, and you feel that patches look "tacky," you can apply them to the inside of a dry suit just as well as to the outside. This may prevent the outside of the suit from looking like a patch quilt, but eventually the wear will show up on those surfaces as well.

Neoprene rubber must be dry when a patch is glued into place. A hair blow dryer can be used to dry out the material if you don't want to spend the time to air dry it.

Patches should be cut large enough to overlap the hole, similar to patching a leak in a tire inner tube. Patches can be cut into square shapes, but octagons (eight sides) are even better. Before the patch is glued down, trim the outer edges of the inside of each patch. This way, the edges will glue down more securely and have less chance of being caught on things and torn off. They will also look better (See figure 5-2).

A thin coat of glue should be applied to both the patch and to the surface of the suit where the patch will be placed. Once the glue has dried to the point of being tacky, a second coat of glue should be applied to each surface and also allowed to dry until it is tacky. An optional third coat of glue may also be applied to each surface in a similar manner, and allowed to become tacky. Then the patch should be applied. The same process works both for gluing neoprene patches on a suit, or adhering inner tube rubber pads using contact cement. Obviously, all

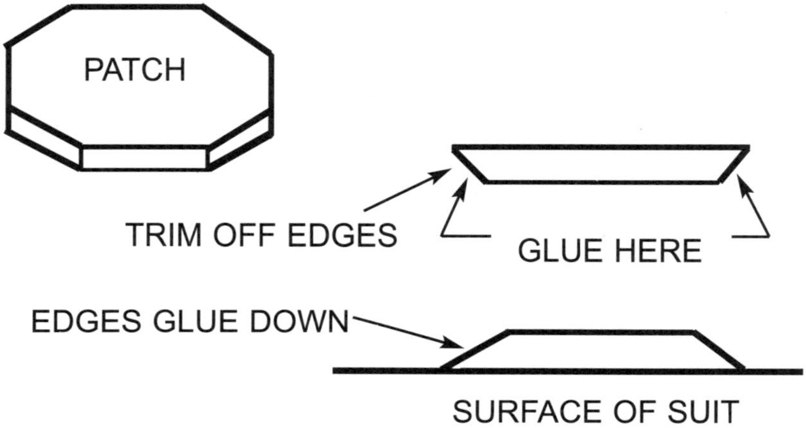

Figure 5-2: eight-sided patches work great - especially, when the inside of the outside edges are trimmed off and glued down

foreign matter must be removed from both surfaces before you begin applying the glue or contact cement.

Occasionally, a newly-installed patch will begin to detach shortly after you place it into service. When this happens, just dry it out again and reglue it, using the same method. It will rarely come off a second time.

After a dry suit has been used extensively, the neck, wrist, and/or ankle seals may become stretched and lose their efficiency in keeping water out. Sometimes, the seal will not fit tight enough, even if the suit is fairly new. This condition can be corrected by cutting a narrow triangular strip out of the seal and then gluing the seal back together again, as demonstrated in figure 5-3. This corrective action will make the seal tighter, and it will do a better job of keeping water out of your suit.

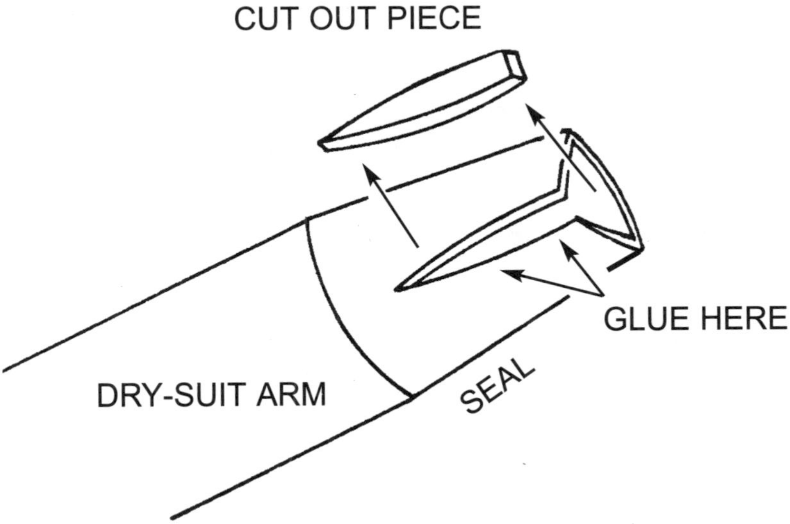

Figure 5-3: Cutting a section out of a seal and gluing it back together to make it fit tighter and seal better.

When a seal is cut and glued back together in this way, it is best to reinforce the new seam with a seam patch. This reinforcing patch will prevent the seam from being pulled apart when the seal is stretched as you get into and out of the suit. On close inspection, you will see that a dry suit seal has a smooth rubber surface on the side which seals against the body. A seam patch should be applied to the opposite side of the seal, so the sealing surface is not affected, as shown in figure 5-4.

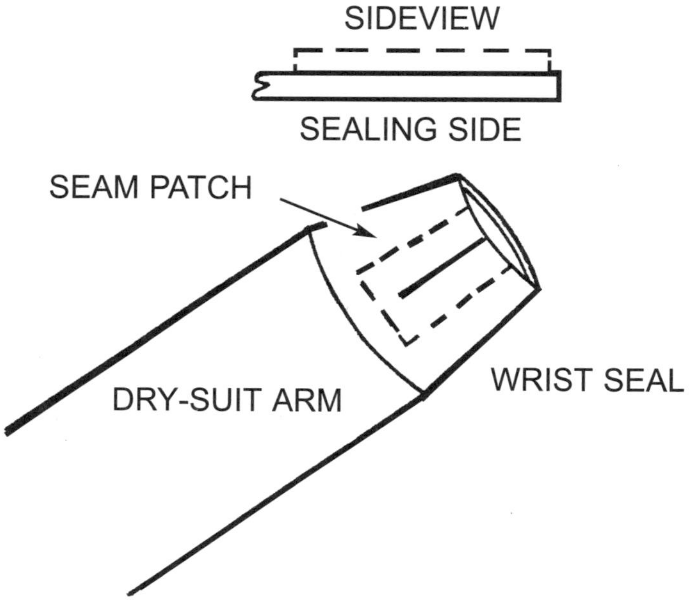

Figure 5-4: A seam patch should be glued to that side of the seal which does not make contact with your body.

Wrist seals take a real beating during dredging activities. They get rubbed and banged around while you are moving cobbles, boulders, the suction hose, and other equipment. Because of this, they may eventually wear out altogether. When they do, you can either send the suit to the manufacturer for repairs, or you can replace the seals yourself.

Replacement seals can be made by carefully cutting the old seals off of the suit, then using them as patterns to cut out new ones from a supply of unused neoprene rubber. The new seals can then be glued and sewed onto the suit. Seam patches in the proper places are a good idea, as noted above. In a similar fashion, if the neck and ankle seals wear out, they can also be replaced.

Caution: Even if the water in the river is freezing, a dry suit can cause heat stroke on a hot day if you are doing a lot of heavy work out of the water. You should be careful to plan your activities accordingly.

HOT-WATER SYSTEMS

Hot-water suits have become very popular with dredgers in the past few years. A steady flow of hot water will make even an old, worn-out wet suit work well in cold water.

As far as I am concerned, if you are going to dredge extensively in cold water, a hot-water system is definitely the way to go!

Some dredgers are using propane, continuous demand, hot-water heaters. But most use heat exchangers mounted to their engine exhaust systems. These systems usually heat water with a heat-exchanging device mounted to the dredge engine. The dredge pump is tapped to provide a water supply, which runs through the heat exchanger, through a steam trap, and then down through a hose, resulting in a constant flow of warm water into the dredger's wet suit.

Hot-water heat exchangers are available on the market. They are also reasonably easy to build. Most homemade exchangers are built with a long length of copper tubing that is either wrapped around the existing exhaust system or is coiled inside a separate housing through which the engine's exhaust is channeled.

The key to a hot-water system is to provide an abundance of hot water. If you do not have plenty of hot water for all of the divers working on a team, then you will likely end up pumping cooler water into each diver's suit--a situation which can be worse than having a wet suit with no hot-water system. You can never have too much hot water, because you don't necessarily have to use it all.

Most ordinary wet suits are adequate as hot-water suits, particularly with the addition of a hooded, shortie vest. Dry suits normally do not make good hot-water suits, unless they are modified. Dry suit seals prevent the hot water from exiting the suit, and will cause the suit to blow up like a balloon. The water trapped inside the suit then cools down and makes the diver cold. Removal of the seals on a dry suit will probably make it a good hot-water suit, as long as it is not too loose. But such use of a dry suit would seem to be a waste of money when a much less expensive wet suit can accomplish the same purpose.

The main problem that dredgers have had with hot-water systems is being burned by extremely hot water or steam. This problem has been solved in recent years by adding a steam trap to the system.

Some prefer to call this a "mixing container." I generally use the Keene plastic air tank for a steam trap on large, hot-water systems. On smaller systems, a mixing container can be made out of PVC plastic pipe.

MIXING CONTAINER/STEAM TRAP

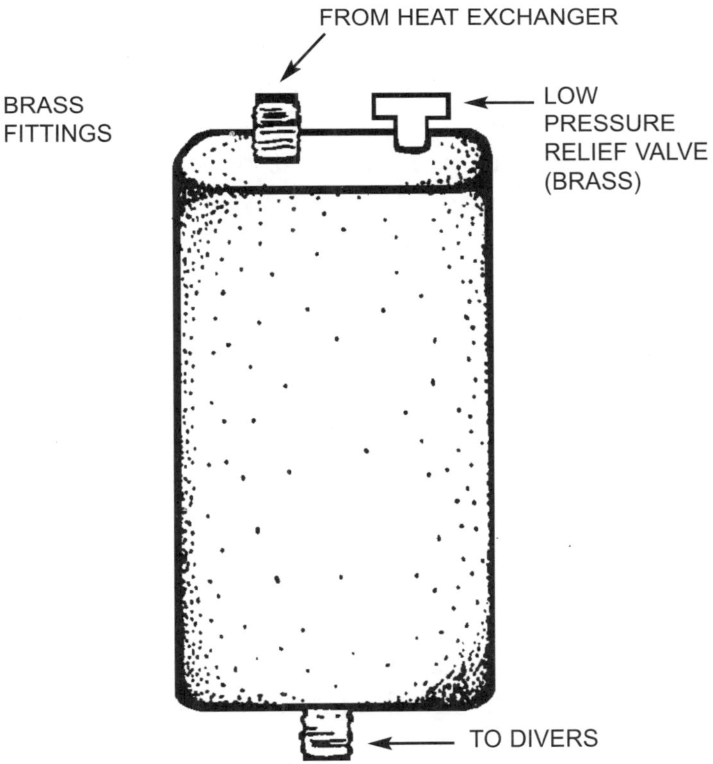

Basically, the mixing container's purpose is to serve as a holding tank for water and steam. If extremely hot water or steam is generated in the heat-exchanger, it will have a chance to mix with the cooler water in the tank before being pumped down to the diver.

The mixing container should be mounted vertically on your dredge with the input entering at the top and the output to the diver coming from the bottom of the container. This way, steam is never allowed to be directed to the diver (it rises to the top). Some systems contain a low-pressure relief valve at the top of the container to allow air and steam to escape.

The mixing container must be large enough to accommodate a shot of extremely hot water, but not be so large that it allows the water to cool down before it is pumped to the diver. Use of this device will allow the diver to feel a rise in water temperature much more slowly. This will permit him/her to remove the hot water hose from the wet suit before it gets uncomfortably hot! Sometimes, the water can be so hot coming out of a heat exchanger, that a special steam hose must be used.

If the water coming from the heat exchanger is too hot to pump directly to a diver (which is often the case), a source of cold water can also be tapped from the pump and directed into the mixing container. By regulating the amount of cold water, you can adjust the temperature of the water being pumped down to the diver. This also increases the amount of hot water that would be available to all of the divers.

Hot water is usually pumped down to the diver through the same kind of hose that we use for air line. The hot-water line and the air line are usually taped together to prevent tangling and additional underwater confusion.

The end of a hot-water line can simply be slipped down into your wet suit from the opening at the neck. Or, you can put together a "splitter" system which will direct some of the water to your chest, hood, each bootie, and each glove. This is the best way to do it when the river water is extremely cold. However, a splitter system can often be avoided simply by having a hot-water system which provides so much volume that the hot water is continually forced into these same extremities.

If you do not have hot water directed to your hands, it may be necessary to use three finger, wet suit mittens to prevent your fingers from going numb when dredging in extremely cold water. But three-finger mittens are bulky. They can prevent you from picking up larger cobbles with one hand, and generally slow you down. However, with a good supply of hot water to the hands, you may find that a set of rubber work gloves will be adequate, as long as rubber bands, or other bindings, are used to prevent cold water from entering at the top.

It will be necessary to warm up your dredge engine to normal operating speed, for at least several minutes, to properly set the temperature of the water that is being directed to the divers from the mixing container.

Generally, if a dredger stands on the deck of the dredge for a few minutes with hot water pouring into the suit, there is no cold water shock at all when entering the cold water. As a matter of fact, it can be a real pleasure to enter the cold water if you happen to run your body temperature up a little bit too much.

A hot-water system should be removed from the dredge when it is not being used during the summer months. Otherwise, the heat and vibration will tend to wear the system out unnecessarily. Also, even when not in use, if a hot-water system is attached, water should be allowed to flow through it any time the engine is running. This will prevent unnecessary overheating of the heat exchanger.

If you are tapping your dredge pump for a supply of water, be sure the water output is either closed or under water when priming your pump. Otherwise, there may be an air leak which can prevent priming.

The nice thing about a hot-water system is that it can supply a continuous feed of hot water into your suit. This way, your body's energy reserves are not being constantly used up to keep you warm. As a result, you can be comfortable and log in more dredging time.

On the other hand, you can also get so warm that your body doesn't want to work, like being too relaxed in a hot shower. The solution to this problem can be controlled by partially opening the zipper on your wet-suit jacket. Believe me, "too warm" is much easier to solve under water than "too cold."

A common question that people often ask is, "Should I get a hot-water system, or a dry suit?" The answer to this lies in what you intend to do. I suggest having both systems available, depending on your activity. For production dredging and sampling in extremely cold water, I would use a hot-water system. For swimming across the river to stretch a rope, or for swimming down the river with mask and snorkel to look at the bottom, I would recommend a dry suit.

OTHER WINTER GEAR

A pair of tall, rubber boots that you can pull on over your booties is a good way to protect your feet, your booties, or your suit below the knees. If you poke some small holes near the bottom of the boots, the water will drain out when you exit the river.

Steel-tipped, rubber boots are great for protecting your toes from falling rocks and rolling boulders. During the winter months, though, steel-tipped boots will make it more difficult to keep your toes warm.

While a full-face mask will help reduce exposure to the cold water, I do not recommend using one unless the water is so cold that you positively cannot get along without it. This is especially true if you are dredging in fast water. When a full-face mask gets swept or bumped off your face, or when it loses its water tight seal, you lose your air supply as well as your vision. Because full-face masks are larger, they are more prone to bump against objects, or be pushed by water currents. When a full-face mask loses its seal, it can be more difficult to clear your mask and regain your composure under water. Since this is a safety issue, you will need to make your own decision.

There are fiberglass diving helmets available that will keep your entire head dry. Since this is commercial equipment, you will need to determine whether the air compressor on your dredge provides enough air pressure to operate the helmet. Sometimes the helmets need to be modified to operate at lower pressures.

Three-finger mittens are better than five-finger, wet suit gloves for keeping your hands warm when dredging in cold water. But, as noted above, you will lose dexterity. You don't want to give up too much of your hand and finger flexibility, because, as a dredger, you will be using your hands a lot.

Inner tube rubber strips can be cut and glued onto the heavier-wear areas of the mittens or gloves to prolong their life.

Heavy-duty, rubber work gloves are excellent for protecting your hands when the water is not too cold. They allow plenty of finger movement, so cobbles can still be easily moved with one hand, while the other hand continues working the nozzle. When the water does start to get cold, rubber work gloves can be extended by gluing on a section of neoprene rubber, as shown in figure 5-5. This extension will prevent a lot of the cold water from circulating into the gloves, and will thereby keep your hands warmer when using them. If the water becomes too cold even for this, you have other remedies available with the hot-water system discussed above.

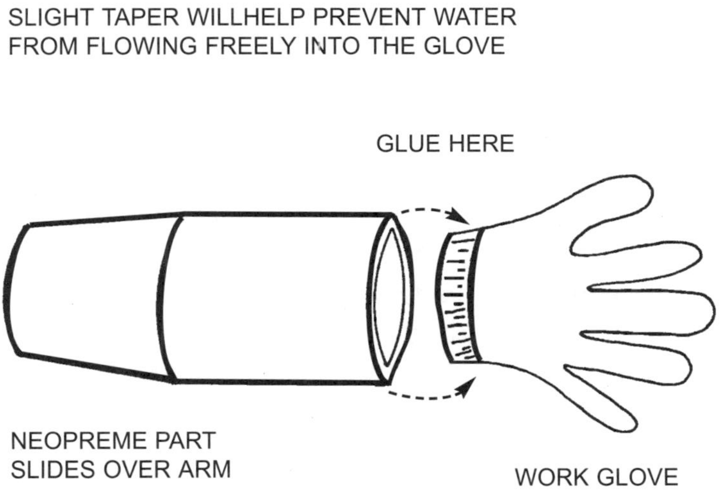

Figure 5-5: Heavy-duty work gloves can be extended with neoprene rubber; so they can be used in colder water, too.

OTHER THINGS TO KNOW ABOUT COLD-WEATHER DREDGING

If you are working in temperatures that are freezing above water, there are certain things that should be done to your dredge each day before calling it quits. Your pump should have a drain plug tapped into the bottom. This way, you can drain the water at the end of each day to prevent your pump from freezing at night. It is not a bad idea to bring some hot water with you everyday in a thermos. Sometimes, the pump will freeze up a little, even with the water drained. Be careful, though, not to crack the pump housing by pouring scalding water directly over it when it is frozen.

Also, in freezing weather, the concentrates and water must be completely emptied out of your recovery system at the end of each day. Otherwise, they will freeze solid and prevent the system from working properly until it thaws out the following day--if it thaws!

Your concentrates from the day should be stored under water or taken to camp, to prevent them from freezing on the bank. Your mask, hood, and gloves should be brought back to camp each evening and kept from freezing.

WINTER DREDGING

The main danger of winter dredging comes from the increased number of storms, and the possibility of losing your dredging equipment when the river water rises up and rages.

While each area will be different, it usually needs to rain for quite a while, maybe a day or two, before the water will rise up to dangerous levels in a river. Sometimes, a cloudburst upstream or up one or more of the tributaries can bring the water level up quickly and unexpectedly on a river. Besides concerns about your dredge in the water, you also have to be careful where your dredging support gear and/or your concentrates are being kept up on the bank.

When you leave your dredge alone, during the

non-production periods, it is good practice to tie it off in at least two separate ways, using two separate lines. The dredge should be tied in such a way that it will be pulled further in towards the bank if the water level comes up. The suction hose should also be tied towards the stream bank forward of the dredge, so that an increase in water movement will not cause the hose to be swept down river and possibly break it loose from your dredge. Miscellaneous support gear should always be stored a safe distance above and away from the water during non-production periods of the winter months.

If your dredge is on the river, pay very close attention if it starts raining and the river begins to rise and turn muddy. As the water level rises, you may need to readjust your dredge's tie-off lines. If the dredge is already located in a precarious position on the river, you should move it before the water level gets to a point where you no longer can move it safely.

If your dredge is set up on the opposite (from the road) side of the river, and you can only reach it by swimming, you should pull it out of the water at an early point and secure it well above the water level. DO NOT wait until the water level becomes so high that you cannot rescue your dredge without endangering your life. The river can get very nasty during a storm!

It generally takes a while for the river to rise up to dangerous levels, though. You should carefully monitor its progress, even if it means setting the alarm clock and going down to check it several times during the night. The important thing is to act well before it is too late to take corrective action. You really cannot be too careful when the alternative is having your dredge washed down river and destroyed. Or, more importantly, having your life placed at risk.

The main danger to your dredge in a storm is not so much the increased volume and velocity of the water, as it is the logs, trees, and other debris being washed down with the flow. These floating perils can snag your dredge or its tie-off lines, and rip it loose from its anchors in a moment. So, when the winter storms roll in, be vigilant and take all necessary and timely corrective actions to protect your dredge and your own personal safety.

SAMPLING DURING THE WINTER

Another fact of life that winter dredgers often have to deal with is higher and faster water. While the higher water will allow you to mine further up the side of the streams, in many areas it will prevent you from mining out in the faster and deeper-water areas, where some of the richer deposits can be found.

On the other hand, if the river edges are paying well, the winter months may be the only time they are available for dredging. The location of deposits will vary from one place to the next. I know of dredging operations that do very well working areas that are normally high and dry during the summer months.

While wet suits, hot-water suits and dry suits do make for good insulation under the water, they generally provide poor insulation to the cold air above water, when you are wet. Therefore, be sure to have a warm winter jacket handy to wear over your diving suit while taking breaks on the surface.

If you are mining in extremely cold conditions, it really helps to have a warm and comfortable camp. A person can put up with some pretty cold and miserable conditions, as long as he knows there's a warm shower and a hot meal coming later that evening. There are few things worse than freezing all day and then going back to stay in a wet and cold camp!

HOW TOUGH ARE YOU?

It takes a pretty tough cookie to dredge in extremely cold conditions. Even with the best equipment, you still have a lot of cold-water exposure on your hands and face. You will spend quite a bit of time working on the dredge, tying off lines, swimming the river, cleaning up concentrates, making repairs, etc. This all adds up to exposure that can be painful or uncomfortable.

Some people are gung ho enough to dredge in extremely cold water on a short-term basis. Few people are willing to do it long term.

We all have the potential to be tough enough to dredge in extremely cold water. What it comes down to, as it always does, is whether we have the desire to be that tough. We all have our limits.

I was mining with several guys in 34° water one winter. One of the divers and I were sampling for a new deposit, while the other two guys were actively dredging up a rich deposit we had already located. They were pulling out several ounces of nugget gold each day, while we were knocking out sample holes. One day, we helped get the production operation started and then headed out to do our sampling. We soon realized we had forgotten our lunch, turned around, and drove back to where our partners were dredging just moments before. We weren't gone five minutes, and they had already gotten out of the water and were in the truck with the heat turned on, drinking hot coffee! They got out of the water as soon as we left. These were tough guys. But that water was cold!

A partner and I were dredging in Alaska, in October, when things started to freeze. We'd had a very good season, but I wanted to put more ounces into my poke before returning home. Ice had already formed on the edges of the river, and my partner had been ready to leave weeks before. I was determined to spend one more week dredging, because the gold was good, and I had plans for what I was going to do with it. One day, with a week to go, I simply could not make my body go under water again. "One, two, three, go!" But, my body refused. It just wasn't worth it any more. My buddy had already started dredging, but I walked over and tugged on his air line. When he surfaced, I asked him if he was ready to go home. We were on the road later that same afternoon, in a warm truck with the heat blasting. In that area of Alaska, three feet of snow fell that night. We were some kind of glad to be on our way!

6

CLEANING UP FINE GOLD

There is an incredible amount of fine gold on many of the gold-bearing rivers. A dredge's recovery system will catch a lot of this gold if it is set up properly.

When you locate, and are producing in, a fine-gold pay streak, at the end of each day your sluice box concentrates will be thoroughly laced with fines and ultra fines. You will find that it is almost impossible to pan off the black sands quickly without losing some of the gold during the process. It can be very time consuming.

There are different kinds of products available on the market that can help you further concentrate your final material. These will reduce the amount of remaining black sands. None of the devices will eliminate all of your black sands while saving all of your gold. Even so, the devices can be very helpful to reduce the amount of black sands you will ultimately need to deal with. I won't go into all the devices in this manual. Different products are continuously arriving and disappearing off the market.

Go check them out for yourself. But, I advise you in advance that no matter what the salesperson tells you, no machine is likely to remove all the black sands while saving all your fine gold, without any loss. So you will most likely end up dialing in one of these devices to work your concentrates down to the least amount of black sand possible, before the fine gold also starts to go. Then you will have to do the remainder of the separation using some other means.

After you have worked your concentrates down to the maximum efficiency that your machines and devices will allow, there are two common ways to do the final separation of remaining black sands from your fine gold. For those who do not care to work with mercury ("amalgamation"), there is a drying and blowing method. This method is thoroughly demonstrated in my video, *"Successful Gold Dredging Made Easy."* It is also covered in my book, *"Gold Mining In The 21st Century."*

Although it takes a little longer than amalgamation, you do not have to mess with chemicals when using the dry & blow method. Many beginners and intermediate dredgers prefer the blowing method for this reason.

AMALGAMATING BY HAND

Amalgamation is the process of using mercury to completely separate gold from the black sands and other heavy elements remaining in a set of concentrates. When done properly, the process is very efficient.

A word of warning: Even though gold dredgers remove large amounts of mercury from America's waterways each year (some existing there naturally, some lost from earlier mining activity), the regulators seem to frown at the thought of miners being unsupervised out in the national forest, while using even small amounts of mercury. Environmental laws and regulations are so extensive and complex, it is almost certain that any use of mercury, anywhere today, falls within the jurisdiction of at least several layers of government. I'm not saying that it is illegal to use small amounts of mercury. But I am saying that it is surely against the law to misuse any amount of mercury. This includes: (1) spilling it on the ground; (2) losing it back into the river; or (3) discarding mercury, or its residues, in any unauthorized place, once it is in your care. I am advising you to understand that as soon as you begin working with mercury or other chemicals, you will have moved your dredging operation into an area of exposure that, potentially, can involve the interest of many different government agencies. So you must be careful how you store, use, and dispose of any of these materials, once they are under your control. Otherwise, in addition to the negative environmental consequences, you may find yourself in more trouble than it was worth.

The purpose here is simply to explain how the mercury process is normally accomplished. It is not to encourage the use of mercury.

One of the most efficient methods of recovering nearly all of the free gold out of the black sands is by amalgamating them in your gold pan. Mercury has a natural, strong affinity for gold. When clean gold comes in contact with mercury, the mercury will naturally adhere to the gold, and hold it to itself, kind of like a magnet. Mercury is a liquid metal. It swallows gold particles into itself. Mercury has no affinity for black sand or most of the other elements found in a set of dredge concentrates.

The following is a common method of amalgamation: First, classify your concentrates through a ¼-inch mesh (#4) screen. The materials that pass through the screen should then be classified again through window screen mesh (#8). Large-sized strainers are available in most hardware or grocery stores, and come in very handy for use in this second screening. (See figure 6-1.)

Once the concentrates are screened twice in this way, the material that remains on top of the ¼-inch screen should be carefully panned (separately) down to just the gold. The reason I say "carefully," is because when panning the larger-sized concentrates, the flakes and small nuggets may not sink very far down into the pan when you shake it. So you have to watch closely to make sure you do not sweep any of them out of your pan along with the valueless materials.

Once the larger of the two classifications of concentrates has been panned to only gold, that gold should be kept separate from the amalgamation process on the smaller classification of materials.

Figure 6-1: A wire screen strainer (large one) is very useful for doing window screen sized classification of concentrates.

The smaller classification should then be panned down to the point where you can see that you will start losing some of the fine gold if you continue on. From there, you should continue your panning over a tub, so the small amount of lost fine gold can be dropped into a container and accumulated for later processing. This "over-the-tub" panning should be continued until you reach the point where you are at risk of losing a significant amount of fines out of your pan.

The remaining concentrates can now be amalgamated, a little at a time, in a gold pan. Simply take a tablespoon or two of those concentrates, and swish them around in the gold pan along with an amount

Figure 6-2: Amalgamating in a gold pan.

of mercury that is approximately equal to, or slightly more than, the estimated amount of gold in the pan. As the mercury picks up the gold, there will be more black sands in the pan which can easily be swept out by panning them into the tub. You use the normal panning technique. It is now easier to discard the black sands, because the mercury in the pan has already removed most of the fine gold from the sands. You vibrate/shake the pan around for a while until the mercury swallows up enough gold. Then, more of the useless black sands can be swept out of the pan. Once a good portion of the black sands have been swept out, you can add another tablespoon of concentrates to the pan. Continue this process until all of the gold has been swallowed up by the mercury from all the concentrates. Add more mercury to the pan if needed.

This process goes much faster if you use smaller amounts at a time, rather than trying to amalgamate an entire pan of concentrates at once.

You want to be careful to not use too much mercury when you are amalgamating in a gold pan. If you do, it will become difficult to keep the mercury from running out of the pan while you are sweeping off the valueless black sands. Mercury is a metallic liquid. It wants to roll around and go with the force of gravity. But it is very heavy. So it will also want to stay at the bottom of the gold pan. The trick is to have enough mercury to gobble up all the gold in your pan, but not so much that it becomes unwieldy.

If you find yourself with too much mercury in your pan, you can pour some off into another plastic or glass container for the time being. You may need to add some back to the pan after the amalgamation process has been underway for a while.

The entire amalgamation process should be performed over a tub of water. This way, if you do sweep some of the mercury and/or amalgam out of your pan, it will be easy for you to recover it back again after you are finished. Once the amalgamation process is complete, the black sands remaining in the tub should be carefully panned off again and checked for any other gold values, amalgam, or mercury.

Once all the gold has been amalgamated out of the black sands, and all the black sands separated from the amalgam, the mercury can be removed from the gold.

HYPODERMIC SYRINGE IN AMALGAMATION

Once you have completed amalgamating a set of concentrates, but before separating the mercury from the gold, all of the excess mercury should be removed from the amalgam. The conventional way of doing this is by squeezing the amalgam inside a thin leather shammy, a piece of nylon, or some other tightly-woven material. A simpler way is to place the opening of a hypodermic syringe (no needle) against the amalgam, and suck the excess mercury out, as shown in figure 6-3.

Some ultra-fine gold will be pulled out of the amalgam when using a syringe in this way. But, this is also true when a shammy or nylon material is being used. It does not amount to much. You can

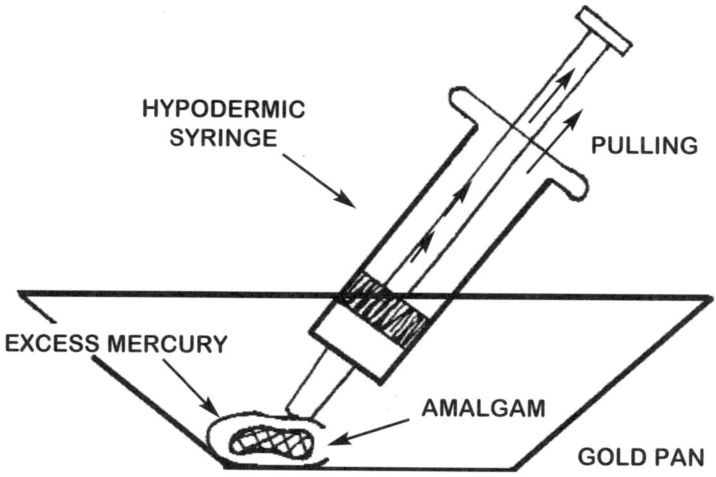

Figure 6-3: Using a hypodermic syringe to suck all of the excess mercury out of some amalgam.

minimize the loss by squeezing the needle hole of the hypodermic (plastic) tightly shut with a pair of pliers. The small amount of gold that may be "lost" into your mercury container can be recovered at a later time by placing that mercury inside the syringe and applying a cotton ball over the opening, as shown in figure 6-4.

You then force the mercury through the cotton, which will trap the fine gold. A solution of nitric acid can then be used to dissolve the trapped impurities, and the cotton can then be dried and burned (be careful not to breathe the fumes) to recover the

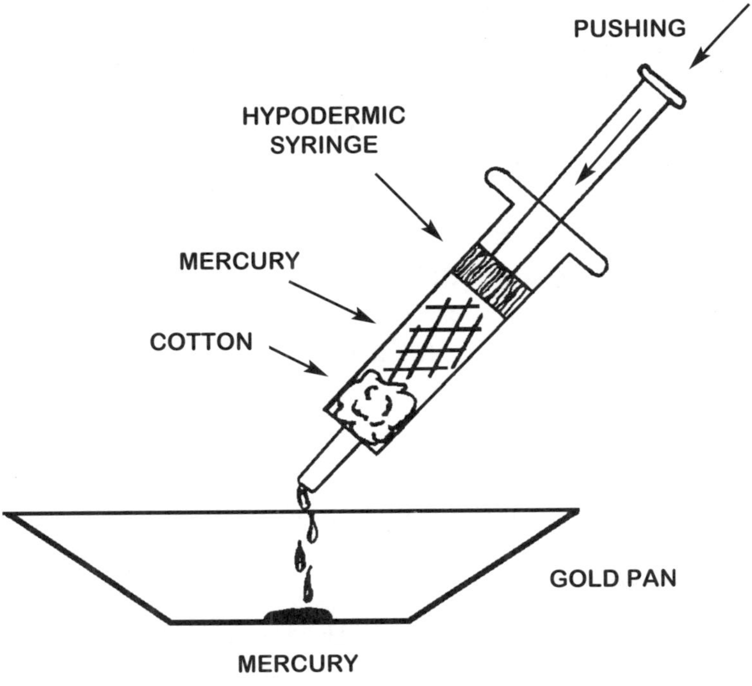

Figure 6-4: Squeezing mercury through cotton to clean it, and to recover all the ultra fine gold.

gold. This, by the way, is an excellent method of cleaning mercury (figure 6-4).

CHEMICAL RETORTING

Once all excess mercury has been extracted from the amalgam, the gold and mercury can be separated either by heat vaporization or nitric acid.

CAUTION: MERCURY VAPORS ARE EXTREMELY DANGEROUS!

They can cause serious illness or even death if enough of them are ingested. Any work with mercury should be done in a well-ventilated area--and, never inside of an air-tight enclosure. Any time you work with mercury, the consequences of your actions must be carefully considered. This is especially true anytime mercury is being heated. If you vaporize mercury, in addition to the concern of breathing the vapors, there is also the danger of exposure to the mercury after it recondenses.

The chemical retorting process can be accomplished as follows: Place the amalgam inside a small glass jar--in a safe, secure location--and downwind of anyone nearby. Pour in a solution of nitric acid (6 parts water to 1 part acid, or stronger). BE CAREFUL NOT TO BREATHE THE FUMES GIVEN OFF BY THE CHEMICAL REACTION! Allow the solution to boil until all of the mercury is dissolved from the gold. Sometimes it's necessary to break up the amalgam with an old screwdriver or piece of metal to expose all of the mercury to the acid solution.

CAUTION: WORKING WITH NITRIC ACID CAN ALSO BE VERY DANGEROUS.

Be extra careful to avoid spilling it onto yourself, splashing it into your eyes, or breathing its fumes. If contact is made, fresh water will help to dilute the acid. Always pour the nitric acid into water when mixing up a solution. This prevents the full strength acid from making contact with impurities and spattering out of the container and onto you or your equipment.

Sometimes it is necessary to add a second dose of nitric acid solution to dissolve the remaining mercury in the jar. If this still will not dissolve all of the mercury, use a stronger solution of nitric acid. BUT BE VERY CAREFUL!

Once all of the mercury is dissolved, pour the acid out of the jar into a second glass jar or other plastic container. A little water can be poured into the jar with the gold, and it can be swished around to pick up the remaining acid. That, too, can be poured into the second jar. You can rinse several times to make sure all the acid solution is removed from the gold.

If you are dealing with a large amount of amalgam, you may see white crystals in the jar with the gold after the acid has dissolved the mercury. These crystals will be heavy. I suspect they are mercury in some kind of crystallized form. You should try to rinse as many of them as possible into the second jar or container, without diluting the acid in the second jar too much. Your success in recovering those crystals will have an effect on how much mercury you will recover when the process is complete.

So now you have the gold in one jar, and the acid solution in another. Drop a small piece of aluminum foil into the acid solution and allow it to sit for a while. A chemical reaction will cause the acid to drop the mercury in the solution to attack the aluminum. This will cause the mercury to revert back to liquid form at the bottom of the jar. It might take a little while for this to happen, though, depending on how diluted the acid solution becomes when you rinse it and the crystals out of the gold jar. Do not use too much aluminum foil. Just use a little at a time until the solution no longer reacts to it. Once the small amounts of aluminum no longer cause a reaction in the solution, and the mercury is at the bottom of your jar, the acid solution can then be rinsed out. Now, you will have your mercury back, or at least most of it.

CAUTION: BE CAREFUL HOW YOU DISPOSE OF THESE ACID SOLUTIONS!

They may be classed as "hazardous waste" by the authorities. Sometimes the local community will have a waste collection program where you can legally dispose of these solutions. Many communities have a periodic waste roundup program where licensed waste removal experts will accept any waste materials that you have accumulated. Before you decide to perform chemical retorting, you must already have a plan for how you will properly discard the resulting waste material.

CLEANING AMALGAMATED GOLD

When mercury is dissolved out of an amalgam mixture with the use of nitric acid, the gold can be left with an unnatural, dark film covering it afterwards. This film can be removed easily by putting the gold into a jar with some vinegar and salt, placing the top securely on the jar, and vigorously shaking it until the gold is clean of the dark film.

If you want to get your gold really clean and shiny, you can rinse out the vinegar and salt, and then follow by vigorously shaking the gold in water with just a little bit of liquid dish soap. Make sure you rinse all of the soap completely away from the gold before drying.

You should be extra careful when cleaning your gold in this way, because it's likely that a lot of the gold will be ultra fine in size. Such fine gold can easily wash out of the jar when the acid, vinegar and/or soap solutions are poured out. The ultra fines should be allowed to settle in the jar first. Watch carefully.

The final gold can then be heated in a small steel pan over a source of heat. CAUTION: Any heating of the gold after an amalgamation process must be heated outside and down wind of habitation. Do not breathe any of the vapors given off.

GETTING THE MOST FOR YOUR FINES

Gold should be thoroughly cleaned and restored to its natural luster before you try to sell it. This is especially true of jewelry gold, when you are hoping to get the best possible price for it.

Smelters prefer fine gold over larger-sized flakes and nuggets. This is because larger-sized pieces of gold have more impurities in them, and will, therefore, have a greater "smelt loss." By this, I mean that when the gold is melted down, the impurities will float to the top and be skimmed off. In this manner, the larger-sized pieces of gold, with their higher proportion of impurities, will experience a greater loss of material during the smelting process than would an equal weight of fines (smaller pieces). This is why it's better to sell your larger flakes and nuggets to the tourists and jewelers, instead of to the smelter. The smelter will give you more money for your fines. The jeweler will give you more for your larger pieces.

7

GETTING YOUR DREDGING EQUIPMENT READY FOR THE SEASON

I always look forward to that time of year when I pull out all my mining gear to get it ready for the dredging season. It gives me pleasure to be back in close contact with it, as I make needed repairs and adjustments. Proper preparation, at this point, helps to avoid problems and delays at the beginning of the season.

My preparations actually begin at the end of the previous season, when I carefully clean all the gear before storing it in a safe, dry place over the winter. Proper "winterizing" includes: (1) running all of the fuel out of the carburetor; (2) emptying the fuel from the gas tank; and (3) spraying some oil into the motor cylinders to make sure the motor does not seize up while in storage.

My "start up" routine at the beginning of each season usually includes the following:

Engine and Pump: I always replace the spark

plug, motor oil and filter, fuel filter, points, condenser, and air filter at the beginning of every season. I also take a hard look at the distributor cap, spark-plug wires, ignition switch and other wiring. Anything that looks questionable is immediately replaced.

Next, I pull the pump impeller to make sure (1) it is not worn or damaged, and (2) that no foreign material is lodged inside. Then, I grease the bearings and check the pump seal, replacing either if they don't look perfect.

Once the motor and pump have been fully serviced, I test them to make sure everything works properly before hauling the dredge down to the river. Note: You should not operate a dredge pump for very long without water passing through the pump. The water being pumped keeps the pump seal from overheating. If your motor test will take very long, you might want to set up a flow of water into the intake side of the pump using a garden hose.

Air-Breathing System: All of the air compressors I have seen on gold dredges have had reed valves. These are stainless steel, flex-plates that allow air to flow only one way through a chamber. These reed valves need to be cleaned, polished, and straightened at the beginning of every season.

Some compressors have rubber diaphragms and some have piston rings. These should be checked and replaced, if necessary. Rebuild kits are generally available for each type of compressor. You can usually pick one up from your local equipment dealer.

I always like to open up my regulators and blow them out with compressed air. Be sure to check to see that there are no holes in the regulator rubber diaphragm, and that the rubber exhaust seal is working properly. This inspection will require you to open the regulator to look inside. If you ever have problems with water getting inside your regulator while you are diving, it is nearly always caused by a worn-out rubber exhaust valve. These are easy to obtain and replace.

It is especially important to open up and clean any check valves in the air system. They have a small spring and ball bearing inside that can get dirty and begin sticking if you don't clean them once in awhile. Note: There should be a check valve on the exhaust side of each compressor, attached to the compressor itself. This is designed to prevent loss of air pressure, through reed-valve leakage, when the motor is shut down or at idle.

I always soak my air lines and regulators in a mild bleach-water solution for about half a day. Let the air line sink in a tub of this solution, so the bleach can kill any fungus that has grown inside the air line. This disinfecting will help you to avoid potential respiratory problems later from the fungus. The line and regulator should be thoroughly rinsed with fresh water afterwards, then blown out with air.

Before placing these into service, I check all of the air lines and fittings to make sure everything is working properly--no leaks and nothing that looks like a potential problem.

I also thoroughly clean the air-reserve tank and make sure it contains no rust, fungus, or anything else that could potentially cause problems.

Dredge: I thoroughly inspect the sluice boxes, screens, riffles, dredge frame, and the jet(s) to make sure there are no holes, cracks, or any other structural problems. If anything even looks like it could become a problem, it should be fixed on the spot. While the gear is in a place where it is easy to fix, all repairs and adjustments should be made **now** to prepare it for the whole season. The goal is to take corrective actions at the beginning, in hopes of avoiding more serious problems after the season starts.

I scrub all of the steel, thoroughly, with a good wire brush and follow up with a fresh coat of primer and paint every spring. It's nice to start off each season looking good!

Be sure to carefully inspect all of your rope and cut out all of the weak areas. Splice or melt all frayed ends, and throw away any old, worn-out rope.

Floatation: I always take a close look at the pontoons to make sure they are structurally sound. If any defects are found, I repair or replace them, as necessary.

Suction and pressure hoses: All of the hoses should be visually inspected carefully before the beginning of the season. It is not uncommon to completely replace all of the pressure hoses. Any holes or kinks in the suction hose must be cut out. If this corrective action makes the suction hose shorter than it should be, I order another one.

Diving Gear: I go through all of my diving gear, replacing anything that is worn out. Wet suits get fresh patches where needed. You don't want to be any colder than necessary during the early spring!

Tools: I go through all the tools, cleaning and oiling them, as necessary. Missing tools are replaced so I can begin the season with a full set. Underwater tools and pry bars get a fresh coat of bright paint, so they can more easily be seen in the work area.

There is a common saying among dredgers that *"Murphy was a miner!"* You know **"Murphy's Law:"** *"Anything that can go wrong, will go wrong."* This statement is especially true for any dredger who starts the season without carefully checking and servicing **all** his gear.

All of this preparation is a lot of work. But, it will give you a solid feeling of confidence, like it does for me, when you crank up your operation on the river each season, knowing that all your equipment has been checked and prepared **exactly the way it should be.**

8

MORE TRICKS OF THE TRADE

Over the years, I have learned a lot of things about dredging. Many of these were learned "the hard way," through unpleasant experiences. By doing things a little differently, or by adding a bit of advance preparation, you can keep your dredging life simpler, with fewer problems and more production. In the hope that I may be able to save you a little time, money, aggravation, and grief, I pass along to you the following "tips:"

1) When moving your dredge downriver through a section of fast water, you should either remove the suction hose from the dredge completely, or keep the suction nozzle well secured during the move. This is especially true if the nozzle has a handle. Handled suction nozzles have a nasty habit of getting hung up on the bottom of the river when being dragged along behind the dredge. Bad things can happen if your nozzle gets caught on the river bottom while you are moving the dredge through

fast water. Among them is **losing** your suction nozzle, especially if you are going through deep or muddy water.

2) If you move your dredge with the suction hose removed, be careful to avoid any hard/severe bumps of the bottom part of the power jet against a rock or a boulder. Such impacts can bend the edge of the jet where the suction hose slips onto it, which can, thereafter, cause lots of extra plug ups.

3) When a new engine is being broken in, it will go through oil faster than an engine that has already been broken in. So, you need to check the oil level of a new dredge more often, to avoid any problems.

4) Dish detergent plastic bottles, with the pop-up tops, work very well as containers for extra engine oil. They keep the oil clean and inside the container where it belongs. The narrow, pop-up tops can make it easier to squeeze in more oil when needed, which can sometimes be a real improvement over the messier method of having to pour it out of a can.

5) Be sure to keep the gasoline and oil containers well away from your face mask. When petroleum/oil products get onto the lens of a face mask, it becomes much more difficult, thereafter, to keep it from fogging up when you work under water.

6) If you accidentally spill gasoline onto your dredge's engine while refueling, be sure to wipe it clean, or allow the gasoline to evaporate, before starting the engine. Otherwise, you could end up with a fire on your dredge. I've seen it happen a

couple of times!

7) One way to put gasoline into your dredge, without spilling it over the engine, is to siphon it out of the gasoline container and into the dredge's gas tank. If you use a clear hose, you can see the gasoline as it comes through the hose and can avoid getting it in your mouth as you get the flow started. Don't use a hose that's too small in diameter, or it will take too long to complete the process.

8) Be sure **not** to gas up your dredge while it is running! That is a great way to set off a fire or an explosion.

9) Here is a good alternative method of siphoning gasoline out of your vehicle, without risking a mouthful of gas. Slide the siphon hose into the tank, submerging it as deeply as possible into the gasoline. Stuff a rag into the gas tank's opening, applying pressure against the rag with your hands, to form an airtight seal inside the tank. When it is airtight, blow into the siphon hose to build up the air pressure in the gas tank. When you stop blowing, get your face out of the way because the pressure inside the tank will start gasoline flowing out through the siphon hose and into your chosen container. If you've ever experienced the taste of gasoline while siphoning, you may prefer this to the "sucking method" of starting a siphon.

10) When dredging in fast water, be extra careful where you put your face mask and work gloves when you surface to take a break. If they fall off the dredge, they will quickly be gone with the current. And, this same rule applies, especially, to your

gold bottle!

If your face mask falls off your dredge, sometimes you can locate it by floating down river, letting the current carry you, while you look for the mask along the bottom. Since face masks are breakable, and can easily become lost, it's a good idea to have an extra one handy.

11) If you intend to swim a rope across a fast-water section of river, and you plan to tie the rope to your body so your hands will be free for swimming, I strongly suggest that you **not** use a slip-tightening type of knot. If you use a knot in which the loop can tighten, and you reach the end of the rope before you reach the other side of the river, the water current can pull you tight against the rope. Then, you may have a **very** difficult time disconnecting from it, especially if it is tied around your chest! This can be particularly dangerous in fast water. I know of one dredger that did this wrong and drowned. If you do tie the rope to your body, I suggest you use a non-tightening loop that you can easily slip out of if you need to.

A "bowline" is a non-slipping type of knot having a loop, and it is easy to tie once you practice it a few times. It is also an easy knot to untie when there is no strain on the line. Figure 8-1 illustrates how a bowline is tied. To tie it, just remember the rule: *"The rabbit comes out of his hole, runs around the tree, and jumps back into his hole again."* (See figure 8-1)

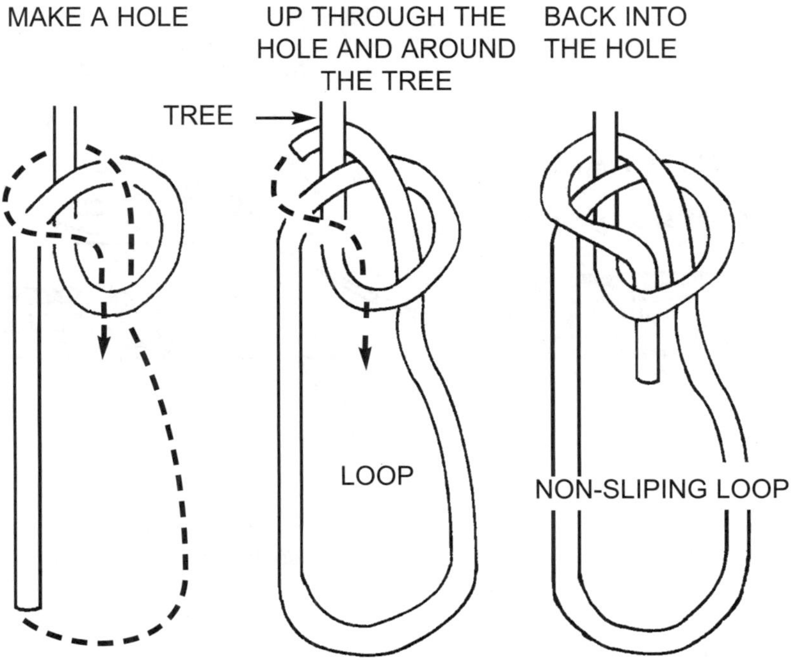

Figure 8-1: How to tie a bowline knot.

12) When crossing fast water with a rope tied to you, consider having your partner **hold** the other end of the rope instead of tying it to some fixed object on the bank. This way, he can easily free you if you do happen to get caught and trapped at the end of the line in fast current.

13) When dredging in fast, shallow water, once you get a hole started, the dredge hole will sometimes cause a slower water area for your dredge to float in more safely, as shown in figure 8-2.

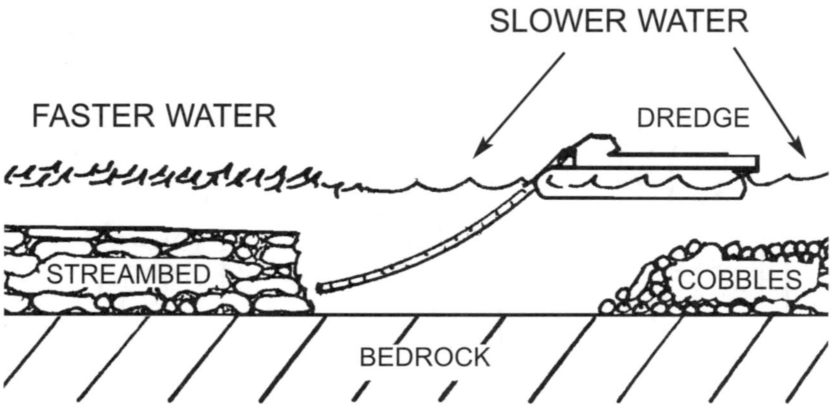

Figure 8-2: In shallow, fast water, the dredge hole will usually slow down the water for the dredge to float in more safely.

14) Never get behind a dredger while he is dredging. He will be tossing rocks, cobbles, and boulders behind him, and you might get bonked on the head!

15) If the water is moving too swiftly through an area that you would like to work, save it until the very end of the dredging season, when the water level drops to its lowest point. You may be able to dredge it safely then.

16) An assay report on placer streambed material has little value in predicting how much gold you will recover with your dredge. Assay reports generally show **all** of the gold values present, even gold that is chemically bonded to the black sands. It will also show those ultra-fine gold values that your recovery system may not be able to hold on to.

The best indication of how much gold you will be able to recover from an area is how much gold is in your sluice box after you have dredged a sample hole there. Assay reports are interesting. But they can also be misleading as to what is actually recoverable with your existing equipment.

17) Don't take too seriously what other people say about a claim owner, or the deals that he/she is supposedly offering. Talk to him/her yourself and make your own deal.

18) Always be considerate of others by keeping your camp clean, especially when you finally leave an area. Watch out for other members of your group who may leave trash and toilet paper lying around. Make them clean up their messes. Don't let them **toss** trash into the river! This is one thing that the regulatory officials, the locals, and the environmentalists all have **good** reason to yell and scream about. There are only a few miners around who are messy. So, take corrective action when you see them do things that will make it harder on all of us.

19) Some dredges use bungee cords to hold the pontoons in place. You may find that one bungee cord on each corner is just not adequate to properly secure the pontoons to the dredge, especially in fast water. In this case, you should double up, using two bungee cords on each corner. You'll find that this gets the job done much better.

20) You can insure your dredging gear against fire, theft, and damage while it is on the river. Make a few phone calls to see which company will give you the best deal. Recreational equipment is usu-

ally covered under a homeowner's policy.

21) If you plan to build a hauling trailer to serve as your base of operations, be sure to set aside a specific area for hanging your wet or dry suits, so they can be allowed to dry out at night.

22) On the type of dredge that uses a plastic primer, you may find that it is hard to prime the pump when there are no concentrates in the sluice box. This happens because, without the weight of the concentrates in the sluice, the dredge will float tilted back towards the engine. This can create an air lock in the primer and prevent priming of the pump. All you need to do is put your weight belt in the sluice box before you prime the pump, and you'll find that it goes much easier.

23) If your dredge engine operates from a 12-volt electrical system, it would be easy for you to set up an underwater stereo system so you could listen to music while you are dredging. Music can help break up the monotony of spending six to eight hours working under water every day. Underwater headphones are available from the suppliers of metal detectors. The speaker wires can be run down to the diver alongside, or inside, the air line.

24) On some types of HOOKA air compressors, you may have trouble with the plastic fan cover vibrating loose and falling off. Extra covers are available in many dredge shops. When you put on a new cover, you can prevent it from vibrating loose by drilling two small holes through the plastic cover, near each screw hole. Then, you can secure the cover to the compressor frame by running a

strand of wire through the drill holes and around each hold-down screw, as shown in figure 8-3.

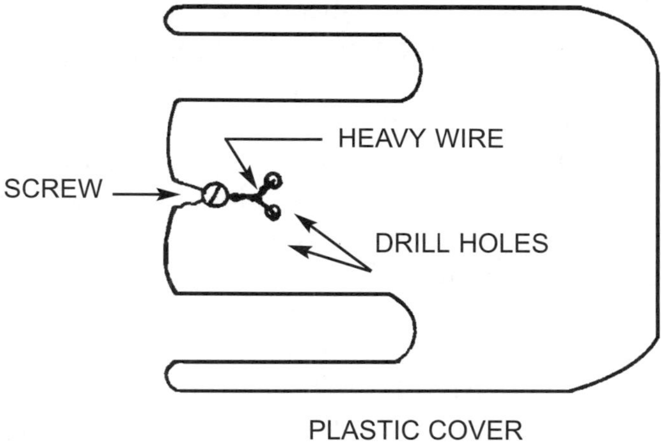

Figure 8-3: Reinforcing the plastic cover on some types of HOOKA compressors.

25) When operating the type of dredge that uses a plastic primer, keep an eye on the primer cap. These wear out occasionally and may allow air to leak through the cap and into the pump when the dredge is operating. Such leaks will cause a loss of suction power at the suction nozzle until the cap is replaced. These caps are usually available at most automotive parts stores. Bring the old one in with you to make sure you get a proper fit for your replacement.

26) If you have an extra weight belt, with at least 25 pounds of lead on it, it will make a great rear anchor for your dredge. This will keep it from

"motor-boating" forward on you when operating in slack water. Be careful to place it far enough behind the dredge so that it is not lost under your tailing pile!

27) After a power jet has been used for a while, it tends to wear out in one particular area, just above the eductor input. Continuous pounding of rocks causes the metal to expand out in that one location. The life of a jet can be considerably extended if you have an additional steel plate welded onto the jet to reinforce this wear area **before** it is allowed to expand out too far (see figure 8-4). When you are having this done, ask the welder to "**tack**" the plate on first, then weld the plate on, a little at a time, moving from one area to another around the plate. Otherwise, if he starts welding at one end and follows the plate around on a continuous weld, the first section of the jet will heat up too much, and the whole thing will warp out of its proper shape.

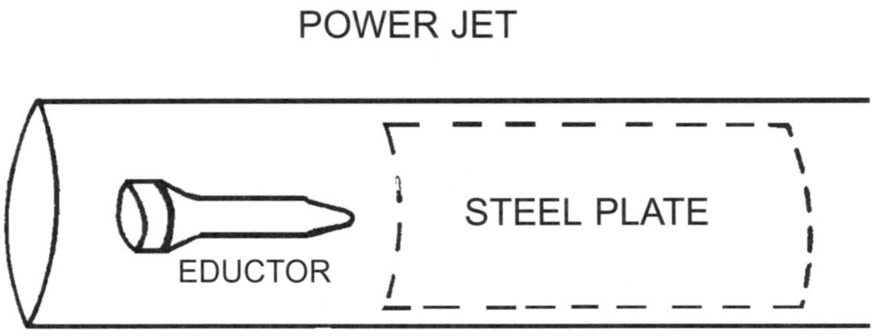

Figure 8-4: Welding a reinforcement plate over the area where a power jet takes the most pounding.

28) Many dredges are manufactured with a suction-type hose attached from the pump output to the power jet. Most suction hose is really not durable enough to last very long at this job. You will find that it starts springing leaks after some use. When the time comes to replace it, use pressure-type hose instead. **Pressure hose** is thicker and much stronger. It is also more expensive, but it will last longer and be more dependable for this type of use. When you buy a new dredge, you should consider paying a little extra, initially, so your dredge can be outfitted with pressure hose, instead of suction hose, for this application.

29) When you're knocking out a plug up and your jam rod slips out of your hand and slides down inside the suction hose, it can be a hassle to retrieve. But, if you weld a T-handle onto the end of your jam rod, it will not slip out of your hand so easily. Also, if the handle is narrow enough to fit inside the power jet, you can use it to locate those really hard-to-find rocks that sometimes lodge in the jet.

30) With the plastic-type of pump primers, sometimes you can put your mouth over the cap opening and suck the water up to prime the pump. You have to be careful, though, to avoid breathing in some water during the process.

31) When assembling your dredge for operation, be careful not to slide the power jet too far into the header box or plastic flare. Reach in and feel its position. It should not slide to the place where the receiver starts to flare out. Otherwise, sand and pebbles (and gold) can get into the flared space

between the jet and the receiver. After hours of vibration, they will become tightly packed and you will find it hard to slide the jet back out of the header box when it is time to disassemble the dredge (see figure 8-5).

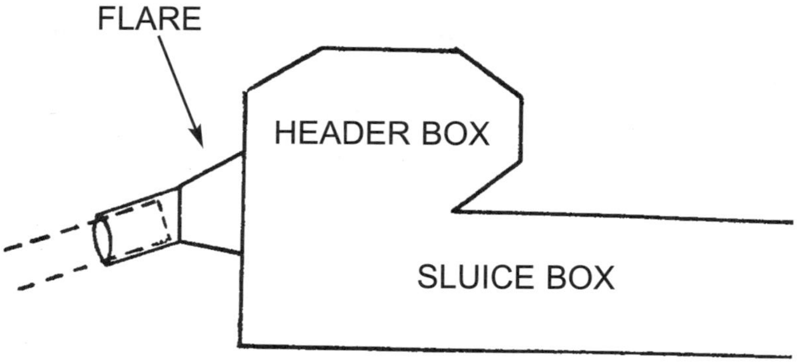

Figure 8-5: Do not slide the power jet into the header box as far as the flared-out section.

32) The plastic, molded-type of pontoons can be pumped full of Styrofoam without adding much weight. This modification will greatly increase their strength and durability. But, be sure to use the kind of foam that does not absorb water!

33) Remote locations should be checked out, thoroughly, with **lots** of sampling **before** you haul in too much gear and commit yourself. Remote areas are more expensive to mine, because you have to bring in all your gas, supplies, and gear. All of

this transportation takes time and effort away from your dredging operation. So, mining in a remote location has to pay much better to make it worthwhile.

Beware of single deposits and small pay streaks. They can fool you into thinking you have really found something great. One or two additional sample holes are all that it takes to tell the difference between a large deposit and a small deposit. Be sure to do this **before** you set up a full base camp in a remote location.

34) When you assemble or build a sluice box, use silicone rubber cement to fill any crack or hole where water could leak out, e.g., where the sluice box bolts onto the header box or flare. If these cracks and holes are not blocked, you will lose gold through them.

35) Avoid setting up your dredge so the sluice box faces (upstream) the river's current. Your tailings can be washed downstream and then be sucked into the dredge's pump intake. This will plug up your pump in short order, and maybe even damage it.

36) If the stream bed is deeper than your dredge's ability to pump material, you can add to its capability by setting up another engine and pump assembly at the surface. This "booster pump" can then provide additional power through a longer pressure hose which can be joined to a second jet, located about halfway down the suction hose, as shown in figure 8-6.

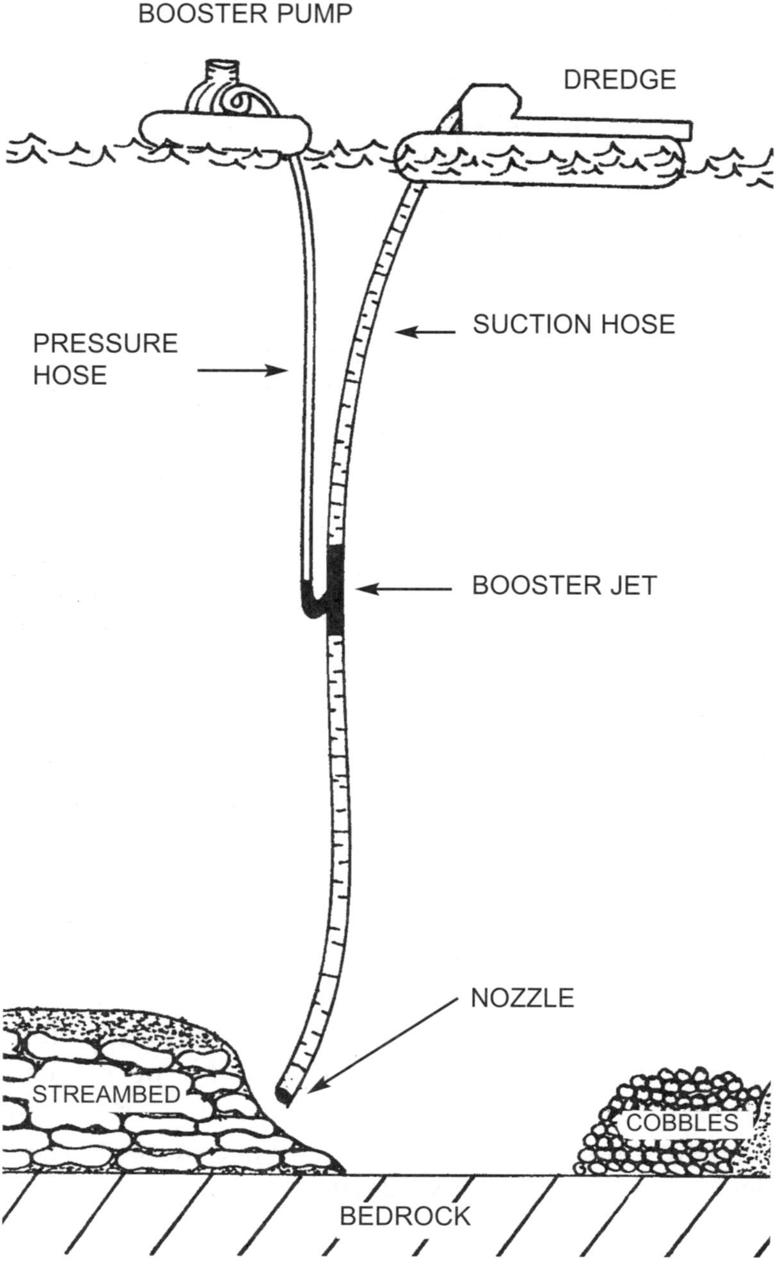

Figure 8-6: Using a booster jet to dredge at greater depth.

37) **Special Concerns When Deep Dredging**.

(a) The deeper you dredge, the more precautions you should take to avoid getting "the bends" (danger increases when deeper than 25 feet).

(b) Since the effects of carbon monoxide poisoning multiply as you descend to greater depths, you have to be absolutely certain that you are breathing clean air.

(c) The deeper you go, the further away from your air source you are. For this reason, you need to make certain that the entire air supply system is trustworthy and will not shut down suddenly when you are down deep.

(d) Your weight belt should be situated so you will have no trouble finding the quick release buckle if you need to bail out quickly.

(e) When making a fast ascent to the surface, from any depth, while breathing from a HOOKA air system, be sure to blow the expanding air from your lungs as you do so. Otherwise, you can rupture a lung in the process. The deeper you dive, the more keenly you need to be aware of this.

(f) When diving deep, you would probably be better off to use SCUBA for an air supply. But, if you are going to use a HOOKA air supply system, you really should have a **very** large air-storage tank at the surface.

In summary, the deeper you are dredging, the more important it is that you observe all safety considerations, including having a diving partner with you. A course in SCUBA diving is essential if you plan to dredge in deep water. You should be **very** familiar with the decompression tables (to prevent "the bends"), and you should know how to take altitude into account.

9

SUCCEEDING AT A GOLD DREDGING VENTURE

Success in gold dredging basically depends upon how well you can master four areas of performance: 1) having the necessary knowledge; 2) having the ability to apply the knowledge; 3) having the willingness to work hard; and 4) having the discipline to manage finances properly.

HAVING THE KNOWLEDGE

This performance area includes **knowing**: (a) how and where to find good, gold-bearing areas; (b) how to operate a dredge; (c) how to sample for deposits; (d) how to dredge up a pay streak; (e) how to recover gold; (f) how to clean and sell gold; and (g) all the other technical "how to" matters relating to a gold-mining program. Mastery of the technical aspects of "how to do it" is very important if you hope to succeed consistently at gold dredging.

The first step in learning something new about a subject is acknowledging that you do not already know everything about it. This admission prepares you for the process of learning more. You must, however, evaluate all incoming information for yourself and reserve final judgment about the validity of what you have learned until you have experienced it in the field.

I have specifically addressed mining and dredging technology in several of my books on these subjects. There are also a lot of other good books available on the market. The knowledge of "how to do it" is readily available from a number of excellent sources. And, having such knowledge in your arsenal of resources is very important to your likelihood of success in the field.

There is no such thing as having too much knowledge. The knowledge you acquire through self-education (books, magazines, videos, etc.) will initially be accepted by you, more of less on faith. But as you apply this knowledge in the field, you will be able to validate it by experience.

APPLICATION

Knowing what to do is only the first step. Beyond that, you must actually apply your knowledge to real life situations. The inability or unwillingness to properly **apply** mining knowledge is one of the principal failures that I have observed in unsuccessful gold dredging operations.

I have talked with dredgers who knew exactly where the sample holes should have been dredged. They could even point out the specific places and give the reasons why. But, for various reasons, the holes were never dredged, or the sampling plan was abandoned before it was completed. Their failures were not necessarily due to a lack of willingness to work. Some of those guys were doing plenty of physical work every day! But, they just did not seem to be able to channel their work energy into those areas that were really essential to get them into gold.

Sometimes, it seems that we are unwilling to depart from some course of action once we get started into it. Sampling requires more emotional flexibility than this. Many times, we are just guessing or hoping when we begin a sampling operation. Then as we sink our sample holes, if we are paying attention, we learn more about the area and adjust our sampling plan accordingly. We learn where the gold path (the highway that gold follows in the waterway) is more likely to be by finding out where it isn't. We find out that the streambed is too deep in some sections of the river by dredging test holes in places where we can't reach the bottom. We have to be prepared to adjust our sample plan each time we learn something new--every time we complete another sample.

A lot of this comes back to identifying and taking the proper approach. This process begins with the fundamental question: *What am I trying to accom-*

plish in the time allowed to me on this project? And the second key question: *Given my available resources, how is the best way for me to accomplish that?* If you answer these questions with integrity, and then direct your activities accordingly, you will be on the right track.

Sampling technology is easy. Gold follows a path in the waterway and deposits at the bottom of hard-packed streambed layers. You find the deposits by first completing sample holes across the waterway to establish **where** (laterally) the gold path is located and in what layer (depth) of the streambed it is found. Then, you sample down to the depth of that layer, along that path, until you find the deposit you are looking for. That is the essence of sampling technology, wrapped up in the last two sentences.

So what makes it difficult to apply? It's the internal emotional struggle involved in not knowing where the gold is. It's easy to get discouraged after working hard to complete several test holes in a given area and then not finding what you are ultimately looking for. Discouragement can override a person's willingness to stick with a sampling plan. People give up on areas that have not yet been properly sampled. Negative emotions can obscure your recognition of positive signs. We all struggle with this. In fact, this is **the** main struggle you have to overcome in gold prospecting. Sometimes, only small traces are the first sign that a gold deposit is near. These small traces of gold are just what we are looking for! We must then follow these traces to see where they lead.

Stop and think about this for a moment: If we already knew where the gold was, we would not have to sample. We could just go into production! But, if it were that easy, someone else would have already taken the gold. The reason there is so much gold remaining in today's waterways is because it is not visible. And, since no one can see it, most people believe it is **not** there.

In view of the fact that you don't know where the gold is when you start sampling (or even know that it is there at all), your real personal challenge is to diligently adhere to your sampling plan, while watching very closely for the signs that will direct your search to those places in the waterway where the deposits are most likely to be found. This, in essence, is the game you are playing in prospecting.

The knowledge about what to look for is secondary to your willingness to apply yourself to the task of recognizing the positive signals when they are there. Your success depends directly upon what you **do** with the information you receive from your samples. These decisions are made internally. To make them correctly requires focus and discipline.

I see people get discouraged when they fail to find a rich gold deposit in the first sample hole. We all experience this to some extent. But, this happens only because we have allowed ourselves to set unreasonably high expectations. The sampling process is never about a single sample hole. Rather, it requires us to spread out across a waterway to find the "gold path," no matter how many sample holes that may take. That's the mission.

Sometimes it takes 4 or 5 samples. Sometimes more. How can you expect to guess it right on the first try? Isn't that an unfair and unrealistic expectation?

It is better to view sampling as a quest to find and follow the gold trail up into the deposits, instead of just dredging a sample hole, or even a series of sample holes. *What are we trying to do?* Find a deposit, right? So, our job is not to simply put down a sample hole. It's about finding and following a trail. This is what you need to be thinking about. This is the wavelength you need to dial in to. Then your focus will be where it belongs.

Don't allow yourself to get too discouraged if a sample hole does not turn up a rich deposit. Instead, feel good about finishing one step in a series of steps needed to track down a gold deposit. Analyze what you have learned from the sample. Consider how this new information should affect the way you proceed in your overall sampling program.

Can you see why personal application is the key to success or failure? No matter how much you know, if you are not out there actively, passionately, diligently sniffing out the path that gold follows, as it lays down deposits in the waterway, you are not going to find very much unless you are just plain lucky. Consistent success depends on applying the right basic approach.

HARD WORK

Gold dredging consists mostly of physical activity. Heavy gear has to be carried around every place you go. Much of your time is spent wrestling with a suction hose, picking up and tossing cobbles as fast as you can, shoving against boulders, fighting to hold your position against the water's current, swimming back and forth across the river, and pulling dredges around on ropes. There just never seems to be an end to the physical work! This is not bad. Unless you don't like hard work.

If you have a distaste for hard, strenuous work, if you don't enjoy it and are always looking for ways to avoid it, you need to find some line of work or form of recreation other than gold dredging. To succeed at gold dredging, you should be willing to take a rather athletic approach toward it, especially during the sampling stages.

Some people are physically inclined by nature, and they enjoy hard work. Other people are not so physically inclined, but they are willing to work hard and do whatever it takes to succeed. Such people can be very successful at gold dredging. But, no matter what your inclination, gold dredging requires **hard work.**

I have found, to be most effective, it is best to attack a gold-dredging operation with a rigid work schedule, just like any other job or business activity. I personally prefer to "pour on the steam" for three straight days. Then, I take one day off from dredging to allow my body to recuperate. The work is physically exhausting on the body if you really pour out the energy. You need to find the appropriate rest interval that works best for you. Otherwise, your body will get overworked and start breaking down. I use my day off to perform gear maintenance and the many other miscellaneous chores that are needed to keep the operation running smoothly. I try get some much-needed free time out of it, as well.

My approach is not the only way. I know successful gold dredgers who prefer to work fewer hours each day, or work at a less intense level of physical activity, but they put in five or six straight days at a time. If we could add up the total units of energy expended on dredging, it would probably come out about the same, either way. It is just a matter of preference and what pace you are most comfortable with. The main point here is that no matter how you cut it, you've got to put the dredging hours in if you want to succeed in mining the waterway.

When people ask me about gold dredging as a profession, I always answer as follows: *"Given the knowledge of how to do it, and the willingness to apply the knowledge, gold dredging is an easy way to make a living, if you are willing to work hard at it."* And, this is the simple truth.

To quickly summarize: (1) to be consistently successful at gold dredging, you should know as much as possible about how to do it (knowledge); (2) you must be able and willing to aggressively apply your knowledge to real life situations in the field (application); and (3) you must be prepared to invest a lot of energy in applying what you know by actually doing it (hard work). If you enthusiastically commit yourself to follow this approach, you **will** get into plenty of gold. There is no doubt about it.

The only element remaining for a successful operation is to manage your recovered gold in such a way that your operation can be sustained, strengthened, and expanded. That concern brings us to the subject of "finance".

FINANCE

This final component goes well beyond dredging, or any physical activity. It has more to do with a person's overall approach on life. The way a person aligns himself to the world is largely reflected by how he or she manages financial affairs and plans for the future. Since a person's fundamental alignment can either reinforce or undermine the other three components discussed above, I'll spend more time discussing this area.

Gold is a very stable measure of wealth. This precious metal has been considered extremely valuable for all of recorded history. Gold is one of the few mediums of exchange that is widely trusted and accepted around the world. No other medium can compare throughout history.

There is security in owning gold or having the ability to acquire it. Never forget that!

Gold can be exchanged directly or indirectly for

food, fuel, shelter, weapons, tools, medicine, entertainment, and most other goods and services that are needed or desired in life.

Normally, in the "working world," one produces a product or service that others need or want enough that they are willing to exchange money for it. In gold dredging, on the other hand, you are able to produce wealth (gold) directly, which can thereafter be exchanged for any needs or wants.

One of the basics of business is that, over time, you **must** make more than you spend. If you consume more than you produce, sooner or later you will have nothing left to consume. Or, you will be consuming someone else's production.

To succeed in gold dredging as a business, you must, over time, recover a greater value in gold than it costs to run your operation. Such costs include all of the expenses related to the operation, i.e., the cost of the equipment, labor, food, fuel, shelter, and repairs.

The amount of gold values that you recover in excess of your mining costs will determine: (1) how much profit you are producing; (2) what the operation's survival potential is; and (3) what living standard you can maintain.

When you are spending more than you are making, you will have to dip into your reserves or borrow to keep operating at that same level. This condition, if it continues, can be dangerous for any business activity. When your reserves are depleted and you still spend more than you make, you will soon exhaust your means of continuing. Cash is the "life blood" of any business. The survival of the operation cannot be sustained for long unless this condition is reversed.

So, one of the primary objectives of any new business, is to become productive enough that you are making more than it costs you to operate. Once you are able to do this consistently and reliably, you can consider your operation to be a viable one and can (somewhat) relax your anxieties about failure. A new business program, however, should **not** relax very much until it reaches this point and develops a confidence that it can stay there.

More than 90% of new businesses fail during their first year of operation. A large number of these fail because they do not have sufficient working capital to get them through the early "start-up" stage, when they are spending more than they are making. It takes a period of time to get most businesses profitably established.

An initial financial investment will be required to set up a gold-dredging operation. Additional operating funds will also be needed to get past the learning curve until you are proficient enough that you can regularly find paying quantities of gold. This might not take that long. But, then again, it would be unfortunate for you to run out of operating funds just as you are beginning to get productive. It is better to plan for a conservative start-up period and begin your commercial program with a large enough reserve to sustain you for as long as is necessary.

When someone is just starting out in commercial

dredging, I always recommend that they be very careful to buy no more than they actually need to operate. Initially, just buy the basic necessities, i.e., the dredge, rope, extra gasoline containers, wash tubs, a couple of gold pans, a good protective suit, and the weights. They should buy nothing more than they actually need to get started. I further advise that they set aside as much money as possible as an operational reserve. I recommend that they initially keep their costs as low as possible, and wait until they start bringing in **plenty** of gold before they begin buying the nice things.

This takes a lot of discipline. Often, if we have the money on hand, we are tempted to buy all the gear and "toys" that we think we **might** need. We want to look and feel good. That's normal. But, if we act on this impulse, we would end up with all this "great stuff", but would not have much in the way of financial reserves to support us while we struggle through the learning curve. Big mistake!

To give you a real life example, I remember a steelworker, from the east coast, who always had the dream of mining gold for a living. When he finally "cut loose" to do it, instead of taking it on as a business, he approached it more like a vacation/fun trip, spending $60,000 during his first year of operating a five-inch dredge. **That was his entire life savings!** He was so busy spending money on *"needed"* things, and living extravagantly, that he actually spent very little time and energy trying to find and recover gold.

Since there were other dredgers in the vicinity

who were pulling out paying quantities of gold, he thought it was going to be easy, once he finally got around to it. The gold, however, **didn't** just come to him. So, he failed to recover even a single ounce of gold during his entire first season!

But, he **did** have all the best, shiny new equipment. When he eventually ran out of his savings, he had to leave dredging and get a "regular" job to support himself.

He could easily have succeeded if he had a more focused approach on finding gold. His focus was wrong. He became a consumer, rather than a prospector. It is **not** the same thing.

I know of another guy who recovered about 20 ounces of gold during his first year of operation. But, he had a different problem. He spent over $80,000 on equipment, paying wages to dredgers, and high-class living. He was just throwing money away! He is out of the business now, too.

Proper financial management is one of the most important aspects of any commercial dredging operation. Money is the "grease" that keeps the wheels rolling. It puts fuel in the gas tank, food on the table, a roof over your head, and it fixes the dredge when it breaks down.

Money can also bring the finer things in life. For that reason, if you do not use **strong** discipline in handling money, it will slip through your fingers faster than you can replace it.

Any individual or organization can spend everything they make, and more, if strong discipline and careful planning are not used. Financial planning and discipline must be practiced in order to keep an operation running and expanding.

One good way to accomplish this is to make a list of the essential expenditures that are required to maintain your operation. Some things on your list may include: gasoline, repairs, equipment upkeep, operational supplies, and basic food necessities. Essential expenditures **must** always be given a first priority in any business. Without them, the operation will stop running and income production will grind to a halt. That would finish your business! So, you must keep the operation running in a healthy condition as your top priority.

Food is a basic necessity. But, it is somewhat flexible in its cost. You can cut substantially on food expenses without slowing down your production. A person can live quite well on rice and beans, if that's all he can afford. Gasoline, however, is a must to run your dredge. So if money is a scarce resource, you don't buy steaks (or even hamburger) if doing so means no gas or some other essential item. That's just the way it is! If you want to succeed, the continuation of the operation must come first.

This is a personal matter of ethical alignment. A person who is grossly mismanaging his or her personal finances will soon be awakened by the sober facts of life. Your success must begin with a personal commitment to make whatever sacrifices are necessary to get you where you want to be.

Once you have your list of the basic necessities, you should make up a second, separate list--one that identifies those things you will need to **expand** the productivity of your operation. Most operations can be improved in different ways. There may be better equipment that can do the job more efficiently, so you can recover more gold. There may be supplementary equipment that can increase your productivity. There may be modifications that will permit your existing equipment to produce better. There may be potentially better mining areas that you would like to sample to assess your opportunities for putting your operation into more gold. All of these pursuits take time and money. None of these things are needed to sustain your existing operation. But they all have the potential to increase your operation's ability to produce more gold (income). Since they are all oriented towards expansion, they go on your "expansion list."

Finally, you might make a third list identifying the items that you would just **like** to have. These may be things that would improve your standard of living, making life more comfortable and pleasant for you.

Living in a small trailer is more comfortable than living in a tent, which, in turn, is nicer than living under a tarp or out of your car. Eating porterhouse steak is preferable to eating hamburger, which is definitely better (for those carnivores among us) than no meat at all. A nice new shiny four-wheel-drive truck is better than an old beat-up truck, which is an improvement over a small car for a dredging operation. All of these things have to do with living standards and preferences. They have very little to do with the capability of your dredging operation to produce wealth/gold. These choices have more to do with wealth consumption than wealth production.

Your ability to prudently manage your finances can make or break any business. The operation will be producing only so much money, so it must be allocated wisely. My own basic formula for success is to shape my financial planning with the following priorities in mind: 1) provide the operation with the basics that are needed to keep it running in a good state of repair; 2) set aside a reserve for emergency contingencies; and 3) direct any remaining surplus toward expanding the operation's ability to increase its gold recovery rate.

Once you have a strong, dependable, commercial dredging program, one that has a healthy reserve set aside, you can then begin to think about directing some of the surplus toward improving your standard of living. But not too much at first. Take it slow.

Contrary to popular belief, throwing money at materialistic desires does little to make the desires go away or bring about lasting happiness. There is **always** the next toy or desirable thing we would like to buy!

On the other hand, there are not many occasions in life that will bring more grief and disappointment than failing at a business activity and losing all your money! Avoiding such a failure is far more important than buying a new car.

If you spend your money on improving your standard of living before you can really afford to, you will reduce the size of your operation and its production/expansion potential. Thus, your standard of living will eventually shrink to less than it was before you tried to increase it.

The main idea is to avoid putting yourself into a standard of living that your operation cannot easily afford to support. It is much better to play it safe than risk the life of your business.

When starting a new business, make a conscientious and determined effort to absolutely minimize your expenses and debts. Make sacrifices in your living standard if you need to, so you can maximize the amount of energy and capital being poured into your operation. Only when it is reliably and consistently producing more income than it is consuming, can you consider draining off its resources to improve your quality of life.

THE NATURE OF COMMERCIAL DREDGING REQUIRES A FINANCIAL RESERVE TO BE SET ASIDE

You could say that the size of a financial reserve has a lot to do with the survival potential of an operation. Every living thing has some natural energy reserve to draw on in time of need. Squirrels hide nuts for the winter. Dogs hide bones for a rainy day. The human body, if it stops eating, can continue going for weeks while feeding off its own energy reserves.

Almost every business activity requires a financial reserve or line of credit.

A commercial dredging operation also should have a healthy reserve set aside. Operations can be temporarily shut down due to equipment failure, an accident or injury, a severe change in the weather or water level, a catastrophic forest fire, or any of a multitude of other reasons. When these things happen (and they **do** happen), if you do not have financial reserves to fall back on, the continued survival of your operation will be at stake.

In any ongoing gold-dredging operation, you will sometimes be into gold and sometimes you will be sampling around looking for it. Since you are usually not bringing in very much gold while you are sampling, your expenses will probably exceed your income during most sampling activity.

Therefore, when you do get into a pay streak, you need to make **a lot** more income than your operation costs to run. It can all average out very nicely for you in the end, if you stick with a good sampling program and work hard.

When you are producing in a gold deposit, you **must** plan financially for when the deposit plays out. Because, when it does, you will have to spend time and money looking for a new pay streak. The need for getting you through this non-income period is why reserves are **second** in your financial planning priorities. These reserves will help insure that your operation can continue in the future. The more reserves that you have set aside, the more secure will be your operation's chances at future survival.

It takes **a lot** of hard work for most of us to get ahead in life. All of your hard work can be for nothing if you don't keep a tight check on how your money is spent and/or if you don't set aside adequate reserves.

Reserves are the result of all your earlier hard work. They should grow over time.

If you have dreams, your desires may seem to be greater than your means to provide for them. This is pretty normal in the consumer-based environment where we live.

The people who **really** get ahead are those who build up reserves and put them to work in low-risk investments that allow the reserves to continue growing. They protect these growing reserves and keep them beyond the reach of their desires to spend, spend, spend!

BORROWING

Try to plan for your dredging venture in such a way that will not require you to borrow money in order to put it on a profitable basis. Debts can put you further into the hole and turn your venture into a losing proposition. Debt service is a burden that should be avoided if at all possible. Onerous debts can be like a heavy chain around your neck. They can trap you, enslave you, and make you feel "owned."

If you get into financial trouble, it is better for you to lower your living standards and to cut off all unnecessary expenditures, so you can pay off your debts as quickly as possible. Borrow only as a last resort. Then, work like a madman to pay it all back as soon as you can. Do a personal assessment to discover how you got into a situation that forced you to borrow. Then take the necessary corrective actions to protect against making the same mistake(s) again.

When I am in need of money, I immediately put my dredge into the most certain gold deposit that I know of. I don't get involved with "long shots" when I'm in that situation. I save the riskier prospects for when I can afford them. When I'm cash-strapped, I go after the "sure thing", and I work harder while I do it.

In gold dredging, as you sample around, you will leave some acceptable pay streaks behind, because richer deposits have opened up elsewhere. Sometimes, you will pick up bits and pieces of information about locations that you know would pay acceptably well. After you have been dredging in a general area for a while, you will pick up leads to a number of such spots. They might not bring in riches; but, they can bring in an income immediately, if you work really hard. You could classify this type of knowledge as a sort of dredging reserve in itself. The longer you are in an area, the more of these "reserves" you will acquire. This is especially true if you belong to an active mining group that has a lot of mining property available to its members.

WHEN YOU ARE INTO GOLD

When you are making a lot of money, the natural tendency is to spend, spend, spend.

If you look around, you will find that experienced moneymakers have managed to overcome this urge. For the most part, successful business people have learned to spend their money on the things that will make them **more** money.

The first thing to do when you find yourself with a surplus of money, **after** you have provided for those things that will insure the continuation of your operation and your immediate reserve, is to pay off all your bills and debts. This complete elimination of debt should always come **before** a standard of living increase. Getting out of debt is a **true** increase in your standard of living, even if you are driving the same old car!

Once you have satisfied all debts, and have a good reserve in place, you should apply any remaining surplus in ways that will increase your ability to produce a larger surplus. This focused approach to resource utilization is how you get ahead in life.

Another caution to keep in mind in commercial dredging: don't buy anything that you cannot afford to fully pay for without endangering the survival of your operation. It is not wise to buy something "on time" that could place an obligation on an uncertain future income stream. If you cannot afford to fully pay for it now, buy a lower-cost version or wait.

I know of a rookie dredger who put his rig into the river and hit an 80-ounce pay streak right off the bat. He was incredibly lucky! He went right out and bought himself a big safe (to keep all his future gold in). He also bought a brand new, four-wheel-drive truck (mostly on credit) and a new dredge. He moved into a nicer, roomier apartment and bought some fancy furniture to fill the extra space. He didn't need **any** of this stuff. It didn't make him any happier in his life. But, it made him feel prosperous to go out and buy it!

After a while, his pay streak ran out, and he didn't have the "know how" to find another one like it. Of course, it didn't take long before he ran short of operating funds. He couldn't make his truck payments, so he started selling off some of his new belongings (at a loss) to try to keep up with his financial obligations. Then, because he was under so much financial stress, he never had the time or opportunity to learn how to find pay streaks. He went bust.

I know of another really nice guy. He went straight to work dredging sample holes along the river and found a small pay streak. It produced a pound of gold during his first few weeks of dredging. Promptly, he went out and paid off all his bills (good move), then spent the balance of his surplus on a down payment for a very nice house trailer (not a good move). Unfortunately, he blew the engine on his dredge the following week. But, he no longer had the money to fix it. So, he put up his dredge as collateral for a loan to repair the engine. When he resumed dredging, he recovered a little more gold, and then the pay streak played out. Because he was also new to dredging, he ran out of operating funds before he was able to find another pay streak. Without the means to continue operating, he couldn't make the payments on his engine repair loan or his trailer. He eventually lost both the trailer and the dredge. That ended his operation, and I never saw him again.

There's much to be learned from these examples. They illustrate that if a surplus is not properly managed, a dredger can find himself in worse shape than if he hadn't found the surplus in the first place!

Surpluses are what you are after in gold dredging. So you need to know from the beginning how to manage a surplus in a way that makes you more productive, not less so, as they are found.

POSITIVE STRESS

All of us have experienced negative stress. Not enough money to pay the bills. Someone important in our life dies or becomes terminally ill. The car breaks down. Accidents. Sicknesses. Getting old and tired. Overweight. Traffic violation. We've all been there. We brace ourselves for the difficulties that naturally happen in our lives. Right? We learn through experience how to deal emotionally with the bad stuff. A whole life of negative worries and events prepares us for dealing with negative stress.

But, consider this: conventional life in our society does **not** prepare us for dealing with sudden bursts of strong, positive stress. "Positive stress" is the affect on your emotions when you are overwhelmed by something **really** good that suddenly happens. Like winning the lottery or finding a buried treasure!

You know the feeling of exhilaration when your favorite sports team makes a great and unexpected score? That is momentary, positive stress.

Longer-term, positive stress can be experienced when you are suddenly faced with a life-changing event of exceptionally good fortune. Most of us don't have much experience with that happening in our lives!

But, the nature of successful gold dredging **requires** that you brace yourself to deal with sudden, positive-stress situations.

I'm not kidding about this. Because, if you really do throw yourself wholeheartedly into gold dredging, the chances are pretty good that you will find yourself uncovering some rich gold deposits. When you do uncover rich gold deposits, the surge of positive emotion you will experience needs to be managed carefully. This is a very personal matter.

The most common term for positive emotion in our industry is *"gold fever."*

My personal-best day in gold dredging was 24 ounces of gold nuggets, taken out of a pocket. The largest nugget was 15.5 ounces. It was about the size of a dollar bill and about 1/2-inch thick! I was dredging with several other guys. We got so excited, we could no longer focus well enough to dredge any more for the rest of th^ day. Using good judgment, I gave the crew the day off and encouraged them to get out of camp. That 24 ounces was the point of discovery of a rich pay streak that produced a strong surplus of income for quite a long time. Financial affluence brings about emotional affluence. Both must be managed.

The best way I can describe the feeling of an overwhelming amount of positive stress, is to liken it to forcing 25 watts of output through a 10-watt speaker. You know how the sound distorts? Positive stress distorts your emotions in a similar way. It is very uncomfortable. Even if it is *enthusiasm* or *exhilaration* that you are feeling. The distortion does not feel good. A person, like a sound speaker, can only handle so much at a time before experiencing overload.

You have to be careful to manage the distortion so that it does not get you in trouble.

A dredging friend of mine came by to see me one afternoon. He was so upset that he could not talk. His hands were shaking so hard, at first I thought he was having a stroke or something. He walked me back to the trunk of his car and showed me 80 ounces of gold that he pulled out of a bedrock crack in one hour of dredging that afternoon. It was all sitting in an unexpected deposit. He was alone. He told me that summoning the discipline required for him to finish cleaning the gold out of that crack was the hardest thing he had ever done. He could not bring himself to uncover the next crack in the bedrock. He was so distorted/overwhelmed, it felt like a drug overdose. He said it was everything he could do to not swallow his regulator.

After leaving me, he went over to show the claim owner, who was entitled to 10% of the recovery. The claim owner got so upset, he kicked my friend off his claim and told him to never come back. That night, 6 commercial dredges on the river were soaked down with gasoline and burned. It's really amazing that no one was killed. I never saw that dredger again. His life was changed forever. The claim owner then got into a battle with his own partners over the claim. None of them ever got to mine a single ounce of that gold. Years later, another dredging friend of mine got in there and pulled out over a thousand ounces of gold nuggets. He somehow managed to keep his crew in line throughout the project. That was unusual. But, he had planned ahead.

I have watched best friendships dissolve over small pay streaks. They were big to the guys dredging them at the time, though.

I remember a 78-year-old, mild-mannered man (who said he didn't even care about gold) who found a rich pay streak using a 3-inch dredge. The next day, he pulled a pistol and threatened another guy who got too close to his gold deposit. The other dredger knew better, but he couldn't resist the temptation. He was trying to move in on the deposit with a 5-inch dredge. He didn't figure the old man would do anything.

I watched a rich deposit destroy an entire family that had hung together bravely during the hard times. Even though they were able to successfully endure the tough times, when **really** good fortune suddenly entered their lives, it proved to be their undoing! They never had a chance. They had never experienced **intense** good fortune before, and did not know how to manage it.

You must be careful not to allow the sudden onset of exceptionally good fortune to ruin your life. That's no joke! Most dredgers usually don't think about this in advance. That's because we're not accustomed to planning for fantastic success before we actually see it. When it happens though, it happens all at once. It blind-sides you. So prepare yourself, and your reaction to this, in advance.

It really wouldn't be right for me to guide you in the ways of recovering gold from rich deposits, without also warning you of the dangerous consequences that can follow enormous success. The outcome can completely undermine your entire life as you know it. I've seen it happen!

If you set yourself up in historically-proven, rich gold country, and apply the sampling techniques that I have described in my books and videos, there is really no limit to the potential recovery that you might find at the bottom of a waterway that has been accumulating Mother Nature's treasure for millennia.

And, that's the honest truth.

Don't get me wrong. I'm not trying to discourage you. No way! We are all in this for the glory and the gold. Go for it!

But, just be aware that if you decide to do it, you

should brace yourself in advance for the possibility that you may find something that will put you through an emotional roller coaster ride that you have never experienced before. Something that could put your passions into overdrive. Virgin gold recoveries that can take you beyond exhilaration.

Of course, I am talking about the extremes here. Short of these extremes, there are few things in life more purely enjoyable than uncovering a high-grade pay streak that you discovered by following through on your own sampling plan. It's especially sweet during those first few moments of discovery. It's such a fantastic experience!

But, it's also wise to be prepared for the emotional possibilities of an activity that can be so captivating, so narcotic, in its affect on you. It can be seductive. Can become more important than just about anything else you are immediately involved with in life. Think about this in advance and keep your sense of balance.

We all have our limits. Don't delude yourself into thinking that gold won't affect you that way. Why do you think you are interested in looking for it? Look deeper!

And, enrich yourself by the experience!

Stay safe, good luck, happy hunting,

Dave Mack

GLOSSARY

ACCEPTABLE REQUIREMENT: The minimum amount of gold that you must recover on an average daily basis, to make a pay streak worth dredging on a production basis.

ACCESSIBILITY: 1) Having the legal right to bring your equipment into a location and mine it. 2) The degree of difficulty involved in getting your equipment, supplies, and crew into a location.

ALASKAN DAMPER: A device, usually made of rubber, that attaches to a sluice box and rides along the surface of the water moving through the box. The "damper" does several things to improve the recovery of the sluice box, including smoothing out the flow and breaking the surface tension of the water as it passes through the box.

AMALGAMATION: The process of using mercury to recover the smaller particles of gold from heavily concentrated material.

ANCIENT MATERIALS: Streambed materials that are either located in a virgin streambed or materials that have been carried from a virgin streambed during a major storm and redeposited elsewhere.

ANCIENT STREAMBED: For our purposes, this would include a natural streambed that has been in place since before the early miners arrived on the scene. Those gravels or an area never mined before.

ASSAY: An analysis of an ore sample, streambed sample, or a gold sample to determine the proportions of gold, silver, platinum, and/or other metals to the amount of waste material present in the sample.

BACK TRACKING: To follow a gold deposit back through the shallow material of a pay streak in an effort to locate the downstream boundary of the deposit.

BACK PRESSURE: The hydraulic forces that cause an area of water, within a waterway, to move backwards against the main flow. A back eddy is an example of a back-pressure area.

BAILING OUT: The act of a diver dropping the weights that hold him under water, so he can quickly get to the surface, usually during an emergency.

BEDROCK PAY STREAK: A pay streak that is lying immediately on top of the bedrock.

BENCHES: Also known as "terrace placers." Sections of old streambed that have been left "high and dry" as the present waterway cut its way down to its current level. Many benches still contain large-paying quantities of gold.

BLEEDING OFF: The natural erosion of bench gravels into the present waterway.

BOIL OUT: The effect of a great force of water being moved over the top of a back-pressure area, when the force of the flow being pulled into the vacuum becomes so turbulent that even the heaviest materials, including gold, will be "boiled out."

BONANZA: A very rich gold deposit, either lode or placer, is often referred to as a "bonanza."

BOULDER: Rocks that are bigger than a large cobble. For our purposes, a boulder is a rock that is large enough that you must use your body weight to move it.

BOULDER-BOUND: A dredge hole with so many boulders at the bottom that it is almost impossible to get a boulder harness around any one of them because they are so tightly packed together.

BOWLINE KNOT: A type of sailor's knot characterized by a non-slipping loop. There are many uses for this kind of knot in a dredging operation because it is easy to tie and untie.

CLASSIFICATION: The process of screening out the larger-sized materials from a body of ore, streambed gravels, or a set of concentrates. Classification is usually accomplished through the use of one or more sizes of mesh screening.

CLASSIFIED MATERIALS: Those materials that have been screened through a certain size of classification screen.

CLASSIFIER: A screening device is often referred to as a "classifier." There is a classifier commonly found at the head of most sluice boxes on gold dredges today.

CLASSIFY: See "CLASSIFICATION."

CLEAN UP: The process of separating the gold from a set of concentrates, then cleaning the gold so it recovers its natural look.

CLEAN UP GEAR: Any equipment that is used in the clean-up process, such as mesh boxes, gold pans, etc.

COBBLES: Rocks that are too large to pass through the suction-hose intake nozzle, but are smaller than a boulder. Cobbles have to be removed from a dredge hole by hand.

COMMON GOLD PATH: Because the specific gravity of gold is six times heavier than most of the other streambed materials being swept down river during a major storm, it tends to follow, and be deposited along, a common path. For the most part, this path tends to hug the inside bends while tracking the shortest route between the bends. The first goal in a sampling program is to locate the common gold path within the waterway.

CREVICE: A split, crack, or open fissure in the surface of bedrock. Crevices that are located along the common gold path are likely to trap gold as streambed material is passed over them during a major flood storm.

CLOUD-OUT: A condition that results when the rock-person moves the wrong cobble in a dredge hole, or when material caves in from the side of the hole, causing silt to wash free and obscure visibility in the working area.

DIRECTIONAL-CHANGE BLOCK: (Winching) A heavy-duty pulley used to winch boulders. "Snatch blocks" are anchored in specific locations so that boulders can be pulled in the desired direction.

DOUBLE-BLOCKING: (Winching boulders) A winching technique performed by attaching a snatch block to the boulder sling, then running the pull cable through that block (pulley) and back to the last directional-change anchor. This type of rigging will double the pulling force that can be exerted against the boulder by the winch and pull cable.

DUMPING OFF: The process of removing concentrates from a sluice box or other recovery system.

DUST: Very fine particles of gold, often referred to as "gold dust." A good recovery system will recover a considerable quantity of "dust," often enough (or more) to pay all of the operating expenses. In the early mining days, a pinch of gold dust was worth about a dollar. Today, the same pinch is worth about 20 to 25 dollars.

EFFECTIVE-SIZE RANGE OF GOLD RECOVERY: Any gravity recovery system will effectively recover gold down to a specific size of particle classification. This recovery system will begin losing more particles of gold when they are smaller in size than this effective-size range.

FAST WATER: A condition in dredging when the water current is so swift that it will forcibly carry your body downstream if you do not take measures to prevent it.

FAST-WATER STABILIZERS: Extra flotation that is attached securely to each side of a dredge to increase its stability when moving it into fast water. Fast-water stabilizers prevent your dredge from being capsized or swamped by the swifter-moving water.

FILLER MATERIAL: Material that works its way down between the major rocks of a streambed as it is forming. Filler material can harden over a period of time and cement the rocks together to some extent.

FINDER'S FEE: A percentage of a gold recovery that is given to someone for his help in locating a paying deposit.

FINE GOLD: Particles of gold that are small enough to pass through a 40-mesh screen.

FINENESS: A measuring system used to indicate the purity of a gold sample or specimen. A specimen having a fineness of .900 would be 90% gold. A fineness of .650 would be equivalent to 65% gold.

FLAT ROCKS: It is not uncommon to find an increase in the percentage of flat rocks in a low/back- pressure area.

FLOOD LAYERS: Separate layers of material in a streambed that were laid down during storms occurring at different times. Each separate layer usually contains its own gold in varying amounts. The gold almost always concentrates at the bottom of a "flood layer."

FLOUR GOLD: Same as gold dust. (See DUST.)

FLOURED MERCURY: Mercury that has been broken down into thousands of tiny droplets.

FORWARD TRACKING: To dredge a narrow path forward (in an upstream direction) towards the upper end of a pay streak, following the line that seems to pay the best, in an effort to find an extension of the pay streak, if it is there.

FREE GOLD: Those particles of gold, in an ore or in a set of concentrates, that are not chemically bonded to the other elements.

FULL FACE MASK: A full-sized diving mask which covers the entire face. Full face masks can be used in cold-water diving to prevent the face from being exposed directly to the cold.

GENERAL LOCATION: The larger, general area where a miner chooses to conduct his mining operation.

GRAIN: 1) A term used to refer to a small particle of gold. 2) A unit of weight in the troy system of weight measurement: 24 grains equal 1 pennyweight. One grain equals 64.8 milligrams.

GRUBSTAKE: The credit extended by a storekeeper, or the money put up by an investor, to outfit a prospector and keep him supplied until he makes a strike. In exchange, the person who funded the program will usually receive an agreed-upon percentage of the gold recovery.

GUT: The river's channel is often referred to as the "gut."

HARD PAN: Where the bottom of a flood layer and the top of the layer just below it meet. The layer just below the flood layer is usually hard-packed enough that the flood was not able to wash it away, leaving it there as a false bedrock. "Hard pan" is the contact zone between streambed layers. It is where gold concentrates if it moved with the upper streambed layer.

HEADER BOX: The forward-most section of some sluice box designs on dredges. The header box slows down the streambed materials and spreads them out evenly over a classifier as they enter the sluice.

HEMATITE: An iron ore (Fe_2O_3) which is reddish-brown in color. Hematite is one of the principal non-magnetic minerals found in black-sand concentrates.

HIGH BENCH: See BENCHES.

HIGH GRADE MATERIAL: The richest portions of streambed contained in a pay streak.

HIGH GRADING: 1) The act of stealing the higher grade gold specimens (or a portion of the gold recovery) from a mining operation, usually by one or more of the workers. 2) Also, the practice of mining only the richest portion of a pay streak.

HOOKA AIR SYSTEM: A breathing-air delivery system, used on most gold dredges, which pumps the air from a compressor on the surface, through an extended air line, to the divers down below.

HOT-WATER SUIT: A protective suit used to keep cold-water divers warm by means of pumping water, which has been heated at the surface, down through a hose and into the suit. With a continuous supply of warm water, even an old wet suit can be used to keep a diver comfortable when dredging in cold water.

HYDRAULIC CLASSIFICATION SYSTEM: A means of classifying streambed materials by a recovery system using water (hydraulics) to wash the materials across classification screens. Double and triple sluices are examples of a hydraulic classification system.

HYDRAULIC MINING: (Hydraulicing) A method of gold mining by which a large volume of water, under pressure, is directed by hoses to dislodge gold-bearing, streambed materials located on the benches and hillsides above the level of the existing waterway. The pressurized water stream erodes the gravel down through sluice boxes, where the gold can be recovered.

INTERMEDIATE: A term used to identify medium-sized dredges. Intermediates usually include dredges of the four or five-inch suction hose size. Intermediates are light enough that they can be moved around without too much trouble. Yet they can also process a large enough volume of material to pay well when used in a rich pay streak.

JET: (Power-jet) A tubular component on a suction gold dredge through which water is pumped into the upper portion of the suction hose, thus creating the vacuum that provides the suction power at the nozzle. The jet is designed so that streambed material can be pumped directly into the recovery system without passing through the water pump.

LAYERS: Over a period of time, different flood storms deposit new streambed material on top of older material. These different deposits are called layers. They can usually be recognized by their different colors and degrees of compactness.

LOCKED-IN VALUES: Valuable minerals (gold) that are chemically attached to other elements within the black sands, or an ore, requiring more involved methods of processing so the values can be released and recovered.

LODE: Veins that contain valuable minerals are referred to as "lodes." Gold, in hard-rock form (lode), is commonly associated with quartz veins that protrude through the general country rock that makes up the earth's crust. Lodes are the original source of placer gold deposits.

LOOSE-PACKED MATERIAL: Material that is not tightly packed together, but is loosely situated. Often, when dredging a dredge hole in loose-packed material, the sides of the hole keep sliding in because there is no structure to hold the material up.

LOOSE PACK: See "LOOSE-PACKED MATERIAL."

LOW GRADE MATERIAL: Streambed material that contains lesser quantities of valuable minerals (gold).

LOW/BACK-PRESSURE AREA: Any area in a waterway where the water force slows down, or reverses direction, due to some sudden change in the bedrock configuration, the presence of some obstruction, a change in the direction of flow, or a widening of the channel. Gold tends to concentrate in the low/back-pressure areas that lie along the common gold path. Low/back-pressure areas become pay streaks when gold is present.

MAGNETITE: (Fe_3O_4) The heavy, magnetic black sands (iron oxide) that are found in the concentrates collected by the recovery system.

MAJOR STORM AND RUNOFF: (Major flood storm) An incredible amount of water force is required to pick up and redeposit the streambed and gold found at the bottom of a waterway. Winter storms, in most cases, do not have the power to do this. Once in a while, however, a storm or snow runoff of a huge magnitude will transport and redeposit the streambeds of an entire area. We call these "major storms and runoffs."

MATERIAL: (Streambed material) A term used to include all of the rocks, sand, gravel, mud, clay, silt, and other sediments that make up a streambed.

MERCURY: ("Quicksilver") A heavy, silvery-colored, liquid (at normal temperatures) metal that has a tremendous affinity for gold, silver and certain other metals. Mercury can be used to collect the fine-gold values from a set of heavy concentrates collected by a recovery system. This procedure is called "amalgamation"

MESH: A means of describing the size of a classification screen by denoting the number of openings contained in a linear inch of screening. For example, a 10-mesh screen will have 10 openings per linear inch and 100 openings per square inch.

NITRIC ACID: $H(NO_3)$ A clear, fuming, highly-corrosive liquid/acid that can be used to clean gold of various impurities.

NITPICKING: A frustratingly slow pace of dredging caused by a dredge hole that is too small, or a nozzle-operator's production focus being too narrow.

NOZZLE PERSON: (Nozzle-man) The dredge operator who controls and operates the suction nozzle is normally referred to as the "nozzle person" or "N/P."

OBSTRUCTION: A rock or group of rocks that cause a plug-up in a suction hose or jet.

OVERBURDEN: The lower grade, streambed material that lies on top of a pay streak. In dredging, the overburden must be removed first, before recovery of the pay streak is possible.

PAY: The amount of gold that a lode or streambed is yielding to a mining activity is often referred to as how much it is "paying."

PAY STREAK: A pay streak is a concentration of gold that has formed in a section of river or stream, usually due to the slackening of the water force in that area during a major flood storm.

PENNYWEIGHT: A unit of the troy system of weight measurement. One pennyweight equals 24 grains, or 1/20th of a troy ounce.

PLACER DEPOSIT: Free gold that has eroded away from its original lode and been swept into a stream of running water will tend to accumulate in certain predictable locations. These accumulations are called "placer deposits." This term also generally refers to any deposit of gold that has been weathered out and removed from its original host rock.

PLATINUM: A family of 6 rare and valuable metals that are usually silvery-white or steel-gray in color. Since platinum is sometimes recovered along with gold (it is heavier than gold, with a specific gravity of 21.45, compared to gold's 19.3), the placer miner should know what it looks like, so he doesn't discard it along with the waste materials from his recovery system.

PLAYED OUT: All gold deposits, whether placer or lode, can be mined to exhaustion. When all significant quantities of gold, or other valuable minerals, have been mined out of a deposit, it is said to have "played out."

PLUGGER POLE: (also called a "jam rod") A long, thin, steel rod that is applied from the surface to tap out the plug ups in the power-jet of a dredge.

PLUG UP: A plug up is an obstruction caused by a rock, or combination of rocks, that lodge(s) in the suction hose or power-jet and prevents the passage of material.

POKE: A miner's container of gold. In the early days of mining, a leather bag was often used to hold a miner's collection (poke) of gold. Today, plastic and glass jars are most commonly used for this purpose.

PRIMER: A device on a suction dredge that connects the water pump to the water. A primer is usually designed to be manually filled with water so the pump can be easily primed from the surface.

PROSPECTING: The process of locating a deposit of gold or some other valuable mineral. Prospecting usually involves testing separate samples of material taken from different locations until an acceptable deposit is identified, at which time mine-development activities (production) can begin.

PUMPING MATERIAL: The action of feeding streambed material into the suction nozzle. The material flows up through the suction hose and is directed across the sluice box, where the gold can later be recovered from the concentrated material.

PUNCHING HOLES: See SAMPLING.

RECOVERY SYSTEM: That part of a mining operation, or a piece of specialized mining equipment, that is designed to recover the gold from the material that is directed through the system. The recovery system on most dredges consists of some hydraulic classification and one or more sluice boxes.

RICH: Streambed material that contains high-grade quantities of gold is often said to be "rich."

ROCK PERSON: (Rock-man) When two dredgers are working together under water, one will control/operate the nozzle (nozzle person), while the other one (the rock-person or R/P) moves rocks out of the way. The rock-person contributes to the dredging operation by removing obstructions and distractions so the nozzle operator can be more productive.

ROLLING HITCH: (Winching) A rolling hitch is set up by slinging a boulder backwards, then running the chain or pull cable over the top of the boulder. This will usually free the boulder up by rolling it

ROYALTY: A payment or commission. It is defined as a fixed percentage of the gross or net recovery. A claim owner usually receives a percentage of what an operation recovers from his claim. A grubstaker or the owner of the dredging equipment may also receive a percentage/royalty.

SAMPLE HOLE: Dredge prospecting usually involves dredging a series of sample holes. The first ones are intended to find the common gold path. Additional sample holes are dredged along the gold path in search of a pay streak.

SAMPLING: Exploratory testing of materials conducted to locate high-grade gravels prior to starting a production-mining operation. This testing is usually accomplished by dredging sample holes. The material from each sample hole is processed through the dredge to determine the gold content. The dredger continues to test sample holes until a sufficient pay streak is located, then production-mining activities are commenced to recover the gold from the deposit.

SCUBA: Self-Contained Underwater Breathing Apparatus (e.g., SCUBA tank, etc.) An underwater air-breathing system (supported by tanks of compressed air) that was used during the early days of suction dredging, but which has since yielded to HOOKA air-breathing systems.

SILVER: A shiny, gray-white, precious metal (specific gravity: 10.5) that is usually found in close association with gold. Silver is seldom found as a "free-standing" metal in placer deposits. It is usually alloyed with the gold recovered from the deposits.

SINGLE DEPOSIT: A single deposit of gold can form anywhere along the path it travels in the river, usually where an obstruction or an irregularity or crevice in bedrock is located. The small amounts of gold in single deposits are part of the trail of gold along the common path that it follows in the waterway.

SLOPE: The term commonly used to refer to the downward gradient of a sluice box or waterway.

SNATCH BLOCKS: (Winching) A heavy-duty pulley that can easily be attached to or detached from a cable.

SPECIFIC GRAVITY: The weight ratio of a volume or mass of any substance compared to the weight of an equal volume of water. The specific gravity of gold is 19.3. This means that any volume of pure gold will weigh 19.3 times as much as an equal volume of pure water.

STREAK: Same as PAY STREAK.

SWEPT BEDROCK: Bedrock that is consistently exposed to a force of water flow that is too great to allow any streambed to form on top of it. It is clean bedrock. There is no material sitting on top of it.

TAIL END: The lower end of a pay streak where the streambed material no longer pays an acceptable recovery of gold values.

TAILINGS: The waste material that has passed through the dredge's recovery system and been discarded.

TAIL RACE: Usually a channel cut into bedrock where tailings were directed after being discarded from the recovery system of a surface-mining operation.

TROY WEIGHT: A system of weight measurement that is commonly used by miners and gold dealers. The most common units of weight within this system are:

24 grains	= 1 pennyweight
20 pennyweight	= 1 troy ounce
12 troy ounces	= 1 troy pound

ULTRA-FINE GOLD: This is a generalized term referring to gold that is very fine in particle size. See DUST.

UPCROPPING: An extension of bedrock (usually tilted on end) that protrudes up and out of the foundation of a streambed. Upcroppings can be the cause of single deposits or pay streaks when they lie along the common gold path.

VALUES: Those minerals/metals that are contained in streambed materials or concentrates that are of value to the gold dredger.

VIRGIN GROUND: An area that has never been mined before.

VIRGIN STREAMBED: A naturally-formed streambed in a location that has never been mined before.

WATER DAM: A barrier in a recovery system, below the classification screen, that slows the water down so that fine-gold recovery may be improved.

Other Books & Videos By Dave McCracken:

Books:

GOLD MINING IN THE 21ST CENTURY
The Complete Book of
Modern Gold Mining Procedures

BASICS OF SUCCESSFUL GOLD DREDGING
How To Find A Lot Of Gold
Using A Suction Dredge

ADVANCED DREDGING TECHNIQUES, PART 1
Finding & Recovering Pay Streaks

Instructional Videos:

MODERN GOLD MINING TECHNIQUES

GOLD DREDGING MADE EASY

COMMERCIAL GOLD DREDGING AND
SAMPLING TECHNIQUES